STO

ACPL ITEM
DISCARDED

AUG 7 '75

Gods of Air
and
Darkness

By the same author

COLONY: EARTH

Gods of Air and Darkness

RICHARD E. MOONEY

STEIN AND DAY/*Publishers*/New York

First published in 1975
Copyright © 1975 by Richard E. Mooney
All rights reserved
Designed by Ed Kaplin
Printed in the United States of America
Stein and Day/*Publishers*/Scarborough House,
Briarcliff Manor, N.Y. 10510

Library of Congress Cataloging in Publication Data

Mooney, Richard E
    Gods of air and darkness.

    1. Interplanetary voyages. 2. History, Ancient.
3. Man, Prehistoric. I. Title.
CB156.M67   001.9′42   75-8926
ISBN 0-8128-1815-6

1866834

For Mary

Contents

Introduction

THIS WORK is a sequel to *Colony: Earth,* wherein we attempted to show, primarily, three things:

• It is possible that man neither evolved nor was divinely created, but arrived here as a colonist from worlds elsewhere in space.

• The civilization these colonists created, which was destroyed in a vast catastrophe that passed into legend as the "Deluge," or "Flood," was remembered as a time of gods and a Golden Age.

• This catastrophe was responsible for the great extinction of life at the end of what we call the Pleistocene, giving rise to the Ice Age theory. We suggested that the Ice Age was not in the *past,* but rather is in the *present.*

These ideas were advanced to suggest explanations for many mysteries in the past: the lack of fossil evidence for evolution; the extermination of the mammoth and other species of animals of the Pleistocene; the existence of areas of knowledge in the past which do not fit with the concept of mankind emerging for the first time from a state of barbarism. They also offered a rationale for the building of structures such as Stonehenge and the pyramids, for which no completely believable explanations had been forthcoming.

In many ways this volume will be more difficult to write than its predecessor. In *Colony: Earth* we dealt to a large extent with physical evidence; in this volume the emphasis will be placed more on intangibles. Evidence for many of the ideas we are seeking to elaborate is almost entirely missing.

Thousands of archaeological sites that may yield valuable information and throw light on many mysteries have been investigated either only scantily or not at all. Traces of vanished cities have been discovered in Death Valley in California and elsewhere in the southwestern United States. Odd artifacts have been discovered, usually quite accidentally, throughout North America, where the existence of a former high civilization has never been suspected.

No thorough archaeological investigation has been carried out at Tiahuanaco in Bolivia, and there are thousands of ancient sites throughout Central and South America awaiting investigation. The great number of stone circles and other megalithic alignments scattered throughout Britain and Western Europe have been almost ignored; many have been destroyed, and it is due almost solely to the efforts of one man (Professor A. Thom) that anything at all is known of them.

Of course, there are exceptions—Stonehenge and the artificial hills of Avebury and Silbury have been carefully preserved as monuments from a former era. Measurements have recently been made at the great megalithic alignment complex at Carnac in France, but it may be many years, if ever, before a systematic survey and mapping project of all the megalithic sites is undertaken. Only then may we discover if there was a link between them or a coordinated mathematical scheme of great precision and importance, which could give us a clue to some of the mysteries of the prehistoric past. No one knows what may be hidden far beneath the hills of Avebury and Silbury, or beneath some of the lesser-known pyramids of Egypt, many of which are covered by sand and half in ruins.

With each year that passes our chances of solving the mysteries of our past diminish; time and the elements corrode the traces, and the bulldozers of our impatient age obliterate or cover up what may be evidence of primary importance. Most major finds of the most unusual kind have been made by pure chance: Gurlt's Cube, that strange steel object found inside a piece of coal, was discovered by accident, as was the bell-shaped jar found encased in stone. Iron nails also have been found inside stone, as was the trace of a screw. Parts of ancient batteries were found covered with dust in the corner of a Baghdad museum, labeled as religious bric-à-brac. Chance may uncover something sensational tomorrow, or we may discover nothing new for years.

In some areas of our investigation we have almost no signposts to show us the way. We shall have to try to analyze mythology in new ways; numerous books have been written on the ancient civilizations and their mythologies, but beyond a certain point the experts are baffled. It may be that as long as we look at things in the traditional way we will never understand and never find the answers.

What is the underlying significance of all the world's religions? Why do they all follow the same pattern, and whence came all our varied gods of the past? Why did they appear to have come into existence at a certain point in time and at no other? Why have not primitive peoples today created new religions like the old?

Why is there always, at the back of men's minds, the vague remembrance of a Golden Age on earth? What is the true significance behind many of our deep subconscious images: the idea of the Great Sacred Egg or the sensation, experienced in dreams, of levitation? Why do many people experience strange happenings in the mind—telepathy, precognition? Are there powers of the mind that are real but hidden, lying dormant?

Flying saucers are not new; reports of them exist as far back as written records go. The first is mentioned on a papyrus in the reign of the Pharaoh Thutmosis III, four thousand years ago, and many references in the Biblical Old Testament can be equated with UFO reports. Are UFOs real, and if so, what are they?

All ancient religions speak of a great disaster in the past and a disaster to come. In Eastern religions world ages are mentioned—cyclic destructions—and the Aztecs in the New World, together with the Maya, were obsessed with the idea of future destructions. Has this significance, and, if so, what could it be?

All these things, and more, we shall examine in this book, and try to fit into a logical framework. Many writers are searching for new answers to old mysteries. Von Däniken suggests that the gods were astronauts from other worlds. Kolosimo suggests a similar approach, but with an additional emphasis on the concept of older advanced terrestrial civilizations. Tomas, in his examination of ancient scientific knowledge in his book *We Are Not the First*, considers the possibility that ancient societies possessed the knowledge we have today.

These ideas are not particularly new: Professor Soddy, in 1909, discussed the possibility of a vanished culture that had advanced knowledge, and Donlevey's *Atlantis* and Churchward's *Mu* contain the concept of vanished, legendary empires. At the present time these ideas are being taken more seriously, as older ideas and explanations fail to satisfy.

However, even the newer concepts still seem to be tied, to a certain extent, to more traditional hypotheses. They fail to explain many of the peculiarities that exist, and tend to introduce factors that are semimystical in nature, such as the idea of secret societies of ancient knowledge or the unreal approach to the Atlantis problem made by Madame Blavatsky and the Theosophists. These ideas tend to bring into disrepute any serious attempt to take a fresh look at the ancient world, and have the unfortunate effect of lumping together the serious, logical investigator's ideas with those of the "lunatic fringe."

We shall try to avoid anything which seems mystical or magical. We shall extrapolate from facts and evidence and interpret religion and mythology in the light of modern scientific discoveries. No matter how bizarre our extrapolations may seem, they are not beyond the bounds of the possibilities inherent in our present scientific knowledge or theories.

1

Life and Intelligence

WE DO NOT KNOW how life began on earth, or how and why the reasoning creature called *Homo sapiens*—literally, "Wise man"—originated. Neither do we know whether or not life exists on other worlds in the universe. Possibly we shall not know that until we are able to travel to those other distant worlds in space, and then perhaps we shall also find a clue to our own origins.

At the present time most reasoning people believe in evolution: the idea postulated by Charles Darwin that life had evolved over many ages, changing to fit different environments and altering itself the better to survive. Darwin's theories have been modified somewhat in the light of discoveries made since his monumental *The Origin of Species* was published, but basically the concepts of "evolution by natural selection" and "survival of the fittest" are the keys of modern thought on the origin and development of life. There are others who believe in a "divine act of creation," and many who believe in evolution feel that behind the workings of life's ceaseless change and growth is a mastermind, a "supreme creator."

There also are persons who are unable to believe either in a transition from the inorganic to the organic in the misty past of earth's beginnings or in divine creation, and who seek answers elsewhere. Alternatively then, we have the Panspermia

Hypothesis, first postulated by Svante Arrhenius in the seventeenth century, which suggested that spores, or seeds of life, drifted from the reaches of space and came to life on the sun-warmed surface of the earth.

A later variant on this idea was that these spores had accidentally, or perhaps deliberately, been seeded here by a spaceship from another world.

Divine creation and evolution both suggest a strictly terrestrial origin, which leads to the assumption that there is something special about the earth as the place where life developed. In recent years, however, it has been suggested that as far as evolution is concerned this pattern could be repeated on many worlds where the conditions are suitable for its development. Going further, alternative biochemistries have been suggested for planets that do not meet the requirements for the development of terrestrial forms of life.

On the other hand, the divine creation theory generally assumes that the earth was picked especially by a creator to be the only abode of life in the universe. This gives the earth special meaning to which it scarcely seems entitled when we consider that our planetary system does not occupy any particularly significant place in the galaxy, and is probably one of many millions of such systems.

The Panspermia Hypothesis in its original form postulated that spores would be driven by the pressure of light toward a planetary body, and this would mean that they would be driven outward from the sun. For spores to have reached the earth, therefore, they would have had to come from nearer the sun or the inner planets, and it is doubtful whether such spores would have survived the high temperatures prevailing in that region before reaching earth.

For spores to have reached the earth from the neighborhood of other stars, the only conceivable way they could have been carried here, whether accidentally or deliberately, is by an object of artificial origin entering this solar system.

However they reached this planet, the first traces of life are reckoned to be a blue-green alga which has been dated as 2.5 billion years old. We are left with a huge gap, until complex life appeared suddenly to manifest its traces in the Palaeozoic period some 600 million years ago, with the different phyla (families) of early mollusks, fishes, insects, and plants already separate and developed. The stages which had led up to these divisions cannot be traced. It seems certain that plant life gained a hold on the land surfaces before life eventually emerged from the seas, and this is easy to understand. Until the plants had converted the primitive atmosphere to an oxidizing one by releasing oxygen into the atmosphere, animal life would have had to remain aquatic and draw its oxygen from the water. Only after sufficient oxygen had been released would animal life be able to survive on the land.

It is curious that the biblical Genesis specifically mentions this *order* of creation, viz:

Genesis 1:9. "And God said 'Let the waters under the heaven be gathered together into one place, and let the dry land appear.' And it was so. God called the dry land earth, and the waters that were gathered together he called seas. And God saw that it was good. And God said, 'Let the earth put forth vegetation, plants yielding seed, and fruit trees bearing fruit in which is their seed, each according to its kind, upon the earth.'"

V. 20. "And God said, 'Let the waters bring forth swarms of living creatures, and let birds fly above the earth across the firmament of the heavens.'"

V. 24. "And God said, 'Let the earth bring forth living creatures according to their kind; cattle and creeping things and beasts of the earth according to their kinds.'"

Surely it was not by chance that the chroniclers of old had the sequence ordered correctly. It seems more likely that they had drawn on knowledge of the correct sequence, knowledge of the same kind that we have acquired.

It is somewhat vague, admittedly; we do not have a wealth of detail—there is no mention of what *kind* of beasts emerged, and in what order. No mention of the early amphibians, or the great reptilian order of dinosaurs who lorded it over the earth during the long period of the Mesozoic.

Here we have a great unsolved mystery. We have divided up the past into three great epochs: the Paleozoic, the period of primitive life and the early amphibia; the Mesozoic, the age of the great reptiles; and the Cenozoic, when the reptiles were replaced by the mammals.

No one knows why the dinosaurs died out and were replaced by mammals. Many theories have been advanced for their decline: fluctuations in temperature, which the dinosaurs, presumably being cold-blooded, were unable to tolerate; the gradual drying up of the swamps in which they are thought to have lived, or radiation from a nearby nova. We simply do not know, and all the answers to the problem which have been suggested so far are a long way from being satisfactory.

We cannot be too sure that the dinosaurs were cold-blooded creatures as are modern reptiles—even some of these live in temperate climates with cold winters, through which they hibernate. Further, it has been found that the dinosaurs lived on every continent and in every part of the world. Their bones have been found in England and Europe, America, Asia, India, the Gobi desert, the tundra of the present Arctic lands, and even in Antarctica. It would appear that the whole planet was warm from pole to pole, and there is no evidence to suggest that all of it became cold, or even partially glaciated, during the Mesozoic.

More recent research on skeletons of the Brontosaurus, the largest of all the dinosaurs, has tended to reverse the earlier opinion that this was a swamp-dwelling creature. It had been thought that this animal spent the greater part of its life in swamps, where the water could more easily support its weight without placing too great a strain on its limbs, and where its long neck would enable it to feed on vegetation from the floor of the swamp.

But further examination of its limb structure, together with footprints outlined in solidified soils, have led to a theory that the Brontosaurus's mode of life actually resembled that of a modern elephant or a giraffe. Its heavy, columnar legs were straight rather than bent, and its foot was shaped like the elephant's. The conclusion has been drawn that the Brontosaurus lived on open or semiopen plains and used its long neck to live off the tops of trees in the way that the elephant uses its trunk or the giraffe its neck.

It is possible that others of the great dinosaurs made a similar environmental adaptation, and that the flesh-eating dinosaurs such as Tyrannosaurus Rex preyed on the more harmless vegetable eaters—a situation similar, in fact, to that obtaining on the plains of East Africa today. As we are aware, these plains are characterized by alternate long dry seasons and rainy seasons. If this were so during the Mesozoic, then periods of dessication would scarcely have led to the extinction of the dinosaurs.

The suggestion of lethal radiation from a nearby nova, or perhaps an outburst of excessive radiation from the sun, also presents difficulties—the main one being that if the dinosaurs were killed by excessive radiation, then all forms of life would have been similarly affected, including the small mammals that existed at the time, and the earth today would be barren and devoid of life.

Furthermore, why was it that only the dinosaurs, of all the kinds of life existing at the time, died out? There were also the first birds, the early mammals, many forms of fish, mollusks, and amphibians in the sea and multitudinous varieties of insect life, many of which have scarcely changed since Paleozoic times.

Also, not all the dinosaurs are extinct. There is one species, the tuotara, which lives in New Zealand and has an ancestry directly linked to the reptilian orders of the Mesozoic. Of course, this is only a small reptile—although many of the dinosaurs of Mesozoic times were quite small, one being as small as a mouse—and it is confined to this one area only, with perhaps the island's isolation being a factor in its survival.

An exceedingly primitive fish, the coelacanth, was supposed to have become extinct many millions of years ago, but more than one of these creatures has been caught in recent years. This is further proof that some species from remote times are still alive, and others may yet be discovered.

It does seem, however, that it is not factors of excessive heat or cold or radiation which were responsible for the death of the reptiles of the Mesozoic, as there is little doubt that they would be capable of surviving comfortably in our warmer climatic zones. There must be another answer to what appears to be a selective death of many ancient species.

Could it be that the great saurians did not die naturally, but were killed off deliberately? This would mean an intelligently directed policy of destruction, presumably undertaken by human beings. The objection would be immediately raised that human beings were not contemporary with the dinosaurs, which died out some 60 million years ago, whereas *Homo sapiens* only appeared, according to our anthropologists, some thirty-five thousand years ago.

Are we sure, however? We do not know how long mankind has been on the earth, although the general opinion is that true man, *Homo sapiens,* has not been here for more than thirty or forty thousand years in his present form. There are, nevertheless, some peculiarities. Consider:

• Dr. Chow Ming Chen, in the Gobi Desert in 1959, found the impression of a ribbed sole on sandstone, reckoned to be millions of years old. Dinosaur footprints have been found in similar sandstone beds.

• A rock carving in the American Southwest, at Havasupai Canyon, Arizona, shows a Mesozoic Brontosaurus. A rock drawing, also from North America, shows a clearly recognizable Stegosaurus, also a Mesozoic saurian.

• A design on pottery discovered at an ancient site at Cocle, in Panama, bears a striking resemblance to a pterodactyl, also stemming from the Age of Reptiles.

Shoeprints, drawings of dinosaurs—the artifacts of man—which could be contemporary with the dinosaurs. What are we to make of this? Were the drawings made from life, on the spot? Was someone walking around in shoes at the same time the dinosaurs were living on the earth?

Either mankind was living here at the same time as the dinosaurs—60 or more million years ago—or the dinosaurs did not live as long ago as we have thought. Perhaps our dating is all wrong.

Or perhaps man was not living here at this remote period, but was visiting this world by spacecraft from other planets.

Were there, perhaps, visitors from a highly advanced civilization elsewhere in space who came here and killed off the dinosaurs? We have not been able to find a good natural cause for the extinction of the dinosaurs. In that case, was their extermination deliberate? Either man was living here at this time and killed off the dinosaurs or he visited from elsewhere for the same purpose.

If man was living here at this remote period, then it is understandable that he destroyed the great reptiles. They would constitute a menace, particularly the large carnivorous varieties, and even the vegetarians would consume an enormous quantity of vegetable foodstuffs useful to man.

If, however, he was a visitor from elsewhere in the universe, why go to such trouble? It would be perfectly understandable only if this world had been selected for colonization by civilized people from elsewhere in space. Under these circumstances it would be logical to eliminate any large and dangerous forms of life.

This point of view may mean that mankind, if not already living here, was visiting this planet many millions of years ago, if our dating of dinosaur bones is correct. In this connection, it is interesting to note the unusual finds of artifacts which could date from the Mesozoic or even earlier. There is the example already mentioned of Galt's Cube, a steel cube found in a bed of coal in Silesia, which must have been there before the coal bed was formed. This would place the object in Carboniferous times, many millions of years earlier than the Mesozoic. And there is that jar encased in rock, which could also be many millions of years old. It is not beyond the bounds of possibility, therefore, that spaceships from other civilizations have been visiting this planet over a period of hundreds of millions of years.

We are faced with another problem when considering the extinction of the dinosaurs, and that is the sudden transition from one pattern of life to another. The end of the supremacy of the great saurians ushered in the age of the mammals, and it has been suggested that with the end of the dinosaurs, the small mammalian forms living at the time could then expand and fill the ecological niches left empty. This answer, too, is vaguely unsatisfactory.

As we have already stated, the first appearance of life on earth was a blue-green alga some 2.5 billion years ago. Suggestions have been made that there may have been earlier and even more primitive unicellular forms preceding this—perhaps 3.5 or even 4 billion years ago—which would make it possible that life appeared soon after the earth was formed. We have also seen that clearly recognizable fossils show up abruptly some 600 million years ago, with the major phyla apparently already established without intermediary stages.

As *Scientific American* of August 1964 says:

"Both the sudden appearance and the remarkable composition of the animal life characteristic of Cambrian times are sometimes explained away or overlooked by biologists. Yet recent paleontological research has made this sudden proliferation of living organisms increasingly difficult for anyone to evade. . . .

"These animals were neither primitive nor generalized in anatomy; they were complex organisms that clearly belonged to the various distinct phyla, or major groups of animals, now classified as metazoan. In fact, they are now known to include representatives of nearly every major phylum that possessed skeletal structures capable of fossilization. . . .

"Yet before the Lower Cambrian there is scarcely a trace of them. The appearance of the Lower Cambrian fauna can reasonably be called a 'sudden' event.

"One can no longer dismiss this event by assuming that all pre-Cambrian rocks have been too greatly altered by time to allow the fossils ancestral to the Cambrian metazoans to be preserved . . . even if all the pre-Cambrian ancestors of the Cambrian metazoans were similarly soft-bodied and therefore rarely preserved, far more abundant traces of their activities should have been found in the pre-Cambrian strata than has proved to be the case. Neither can the general failure to find pre-Cambrian animal fossils be charged to any lack of trying."

We observe a further point in a book, *Synthetic Speciation,* by Dr. Herbert Nilsson, professor of Botany at the University of Lund, Sweden.

"If we look at the peculiar main groups of the fossil flora, it is quite striking that at definite intervals of geological time they are all at once and quite suddenly there, and, moreover, in full bloom in all their manifold forms. And it is quite as surprising that after a time which is to be measured not only in millions, but in tens of millions of years, they disappear equally suddenly. Furthermore, at the end of their existence they do not change into forms which are transitional toward the main types of the next period; such are entirely lacking."

The same thing which holds true of plant life also holds true for animal life on earth. Entire groups hold sway for many millions of years, abruptly disappear, and are suddenly replaced by other forms completely different.

This factor is one of the main stumblingblocks to a complete acceptance of the evolutionary theory, which is today being increasingly questioned. It is known that all forms of life, whether they be plant, insect, or animal, are basically composed of elements which are common in the universe. Oxygen, hydrogen, iron, copper, calcium, which are some of the elements that make up living forms, are common in the composition of stars and planets, and even in interstellar dust and gas.

What makes the great difference between organic and inorganic is its *organization.* There is another factor to be taken into account: *something* which defies the inorganic world's tendency to simplification, to breaking down into constituent elements, and instead replaces it with an ability to grow and replicate. It is this argument, now being revived in some circles, that there may be another form of energy, a "life force," which somewhat strengthens the hand of the adherents of divine creation against the evolutionists.

A hypothesis to be proposed   admittedly does not solve the problem of how life may have arisen in the *first* instance in the universe, which is something we may never know, and it will be said that placing the *origin* of life as a whole on some hypothetical planet perhaps many millions of light-years away is dodging the issue. So it is, but so also is divine creation, and the stubborn insistence on an unproven theory of evolution.

What we are attempting to do is simply to advance an alternative suggestion, in the light of the present evidence.

If this planet was visited by intelligent life from other worlds in the remote past, could some of these visits have taken place *before* there was any life here at all? Perhaps this solar system was surveyed soon after its formation, thousands of millions of years ago, and it was observed that at least one planet—earth—was at the correct orbital distance from the primary to support a carbon-based life form. Two other planets, Mars and Venus (and it is possible that there may have been, at this time, a further planet), also fall within what the astronomers term the "life zone," with perhaps the earth as first choice.

Our hypothetical spaceship lands and its crew begins the task of seeding the ground with the vegetation that will start the process of converting the atmosphere from a reducing one to an oxidizing one. A selection of animal forms are introduced into the warm, chemically rich seas. The starship, its task completed for the time being, leaves for home. Perhaps monitoring devices are left behind to register the progress of the early experiments.

In the course of time, when the level of oxygen in the atmosphere has reached a certain level, further forms of life are deposited—those capable of living on the land surface. Larger and more advanced forms are introduced from time to time. We have to assume that this process would take perhaps millions of years, and we are unable to imagine a race which could create a civilization that would remain stable long enough to carry out such a task.

Possibly such projects are undertaken by several races, and it may be that there exists in some superior galactic civilization a central information pool—a great computer complex that contains information on potentially habitable systems where such projects have been initiated, available to any race capable of using the information.

It must be realized that no planet, including the earth, will always be habitable for man, as stars have a finite lifetime and the planets which surround them live or die at the dictate of the primary. The giver of life is also the bringer, eventually, of death. Yet there must be millions of planets in this galaxy alone which have been, are now, or will be habitable for creatures such as man.

Homo sapiens may have originally risen on a world not only millions of light-years away, but millions of years ago in time. If the human race came here as migrants at some time in the past, it may be that it had to forsake a world no longer habitable, or that it was sent here to inhabit a planet made suitable for the race. One can visualize humanity endlessly moving from planet to planet as older ones die and new worlds are born, a continuous restless star-traveling to ensure the continuity of the race. Possibly this process has gone on for countless millions of years, and will do so for millions of years to come.

Eventually, of course, it will be necessary to evacuate a large number of this world's inhabitants, perhaps to the planets of other stars. It may even become necessary, in the distant future, to modify those other planets to make them suitable for human occupation—a concept known as planetary engineering already under serious consideration by our space scientists. It may also be possible to alter human beings by processes of advanced biological engineering to adapt them to environments different from those we are used to.

If we take the concept that man may be a migrant to this planet at some time in the past as reasonable and logical, it would explain better certain things which have been given other explanations. For example, man suffers from an extremely painful spinal affliction known in common parlance as a "slipped disk." Some biologists say this is due to the fact that, as descendents of quadrupedal animals, we are not yet used to walking upright and that too great a strain is placed on the spinal column in what amounts to an unnatural stance.

It could also be, of course, that we are not yet fully adapted to the heavier gravity field of this rather massive planet, particularly if our ancestral home had been a planet with a lesser gravity or if the race had been spaceborne for a long period and become used to a lesser artificial gravity. The fact that the gravitational pull of the earth makes some activities difficult and arduous—climbing and lifting weights—also may indicate that we are not yet fully adapted. This could lead us to suspect that we have not been here for too long a period of time.

If, then, at a remote period, the earth had been prepared for later human colonization, how do we account for the existence of the dinosaurs, which would have to be eliminated at a later stage to allow the settlement of human beings? Perhaps they were an experiment that went wrong—they grew too large, or multiplied too rapidly, or perhaps for some reason or other the earth was left, forgotten, for millions of years. Perhaps the alien biologists were conducting a long-term test with various forms of life under conditions which at that time may not have been suitable for humans, when the sun was much younger.

At any rate, the dinosaurs were eventually replaced by mammalian forms, and a balanced ecological cycle was created, suitable for human colonization at a later stage.

Admittedly, it may sound like fantasy, but in some respects it *does* fit the evidence we have available, and it is not utterly impossible.

Consider: we have life in organized and distinct groups, which appear suddenly and as suddenly and mysteriously vanish and are replaced by other forms, fully developed, which bear no relation to the previous groups. We can say the same thing about mankind, for in spite of the vast amount of research that has been undertaken, no links have been proven between certain extinct anthropoids and true human beings. Mankind appeared as suddenly as the other species and as fully formed.

Many mythologies tell of a God (in the case of the Hebrews) or of gods who created all life on earth. It is perhaps significant that these mythologies tell of experiments and errors that were made by the gods, whereby some of the forms of life they created had to be destroyed. These could be oblique references to the disappearance of forms of life now extinct, such as the dinosaurs. The pattern of life which has been followed—the step-by-step sequence which our biologists tell us is the only logical one and which is supported by religious literature—would seem to have had an intelligent direction behind it. It was neither random nor haphazard. Of course, this is true of the whole of creation. A random or haphazard system would be unworkable, so an orderly pattern would be necessary even if the whole system of the universe were somehow spontaneous and self-creating.

But what if this story of the gods creating life were true? Were they really gods, or have the legends come down to us in a distorted form? Perhaps the deity who created life on earth was in reality a man, or, rather, a team of people—biologists, ecologists, mathematicians. Perhaps there is a hint of this in Genesis where God says: "And let us make man in our image." Why the plural, if there is only one omnipresent God?

We come across this plurality on several occasions. There are the Sons of God who mate with the women of earth and produce remarkable children. Does this perhaps relate to visits by astronauts at later times, who had children by the descendants of the earlier colonists? One God who created all the universe surely would not have children. It has been explained by some experts that the Sons of God was merely a title given to a certain group—perhaps a religious sect like the Essenes—but such people would hardly have produced extraordinary children like "the mighty men of old," as it says in Genesis. It would also appear that this part of the Genesis mythos stems from an extremely remote period, to judge from its lack of detail.

In Isaiah it is stated that God was coming in his great anger to destroy the whole land, and that he came with a mighty host and the weapons of his indignation. Once again, the hint that "God" was not singular but one of many.

We could look at this problem from the other end of the scale and imagine that man has reached one of the inner planets, say, Venus or Mars. If soil conditions on Mars are not too hostile, it may be possible to introduce extremely hardy alpine-type plants, such as high-altitude grasses from the Andes or the Himalayas. If they flourished in sufficient quantities they would, in the course of time through the action of photosynthesis, release oxygen into the Martian atmosphere. A denser, oxygenated atmosphere would mean that the planet would retain more solar heat, and the planetary temperatures would rise.

At a later stage lower-altitude plants could be spread, and eventually a stage would be reached where plants, including trees, could be introduced from temperate or warm terrestrial zones. Once this was accomplished animals, birds, and insects could be introduced, and a balanced ecology of flora and fauna would eventually be reached.

We would now have reached the stage, after a long period of time, when colonists could be introduced to a planet that resembled earth. The atmosphere may still be thinner than that of earth, and the average temperature much cooler. Mars is 141 million miles from the sun compared to earth's 93 million miles. Perhaps the first colonists could be people used to high-altitude conditions—Indians from the Andes or Himalayas, together with Europeans adapted to high altitudes and cold climates.

Later still, settlers from lower altitudes and warmer climates would find little difficulty adapting to their new world. These colonists would build cities and transportation systems, farm the land, undertake forestry and mining, and manufacture all their requirements; in short, they would do everything that we do today on earth. Many generations would pass, and those born on the planet would be perfectly adapted to the prevailing conditions.

The converse is also true: these people, born on another planet, would have difficulty surviving on earth, for the gravity on Mars is only a third of earth's. They would therefore find the gravity of our planet an insupportably crushing weight. People born in a lighter gravitational field would produce adaptive differences; they would be perhaps much lighter and taller in build than their earthly ancestors, and, although more delicate-looking than their cousins from earth, on Mars they would be capable of the same physical feats as earthbound people.

If we imagine that, for some reason, communication was lost between the planets and that this lack of contact was continuous over a period of many thousands of years, and if we further consider that it is possible that future generations of earth-descended Martians may forget their true origins, then a peculiar situation arises.

They may suffer setbacks in their world; disastrous wars could bring a temporary return to barbaric conditions. The ships that transported them would have long before been broken up and their materials used for other purposes, or they may have been returned to earth. After many thousands of years new religions may spring up, and new mythologies be created about their origins.

Eventually they could reach the stage we have now reached. If they had forgotten their true beginnings their researchers would be faced with the puzzle as to where life had sprung from so suddenly and without preliminary stages. They would be able to say quite confidently that the first stage was vegetation, followed by various kinds of animals, birds, and insects, all in their distinct groups, but they would be unable to say how they had evolved there, as the preliminary stages leading to such evolution would be missing.

They would be in the same difficult situation regarding the Martian people. Their anthropologists would seek in vain for the primitive stages which led to the present advanced Martian humanity. Perhaps they would propose an evolutionary theory, even though the main proofs for such a theory were entirely lacking.

Their theologians and philosophers would also have to endeavor to unravel a curious mythology stemming from their very earliest days, if their true origins were not entirely forgotten but were preserved in a very garbled form. Perhaps they had legends about the time when gods came down out of the sky and brought life to a dead world, and when this was done they also put people there, to multiply and flourish. Perhaps there would be a legend about the first people—or were they gods?—who traveled across a great darkness to their new world and had memories of another world which could be seen from the Martian sky.

How many of these legends would be believed by the hardheaded Martian savants? No doubt such "realists" would scorn them as fairy stories from the infancy of the Martian race, myths of a fanciful explanation of how everything got there. Yet they would look in vain for a native Martian origin, and this mythology would actually be the truth of how everything happened.

I believe we today are in the same position as those hypothetical latter-day Martians. The circumstances they would find are *the same that we find outselves in today.*

The foregoing is not just science-fiction fantasy. We have already seen that such planetary engineering may one day be a possibility for our race. Seeding the carbon-dioxide clouds of Venus with bacteria to assist in the converting of the atmosphere to an oxidizing one is being given serious consideration by our scientists, as is the possibility that we may be able to give life to Mars by the methods we have outlined.

Of course, these are extremely long-term projects, and it is doubtful if they could be undertaken in our present economic or political climate, but we may be driven to these actions by harsh necessity in the years to come: perhaps by circumstances external to the earth, or, more likely, by pressures that will be created here by overpopulation, the scarcity of various raw materials and fuels, or, at worst, by a nuclear holocaust.

It seems we ought to see if there is any justification for the concept that we may have descended from voyagers from other stars in the distant past—that we are, in fact, a race whose ancestors were not terrestrial man but galactic man. There are certain elements in mythology that could lead to this concept. The very existence of such myths should make us ponder.

2

Humanity—
A Spatial Origin?

"Mankind evolved from a remote, apelike ancestor which we cannot exactly trace, and which died out after giving birth to *Homo sapiens.*" This, briefly, is what the anthropologists tell us. "Mankind lived in a very primitive state as a hunter and collector of berries and roots for many thousands of years, perhaps even tens of thousands of years, before he 'accidentally' managed to grow some grains and thus founded the first settled agricultural communities, which led eventually to cities and civilization." This is what the prehistorian tells us.

If everything was so simple and straightforward, why are we confronted with so many puzzles about our past? Why are there such huge and all-important gaps in the evidence? Why do we have such curious mythologies? If man had always been primitive and gradually fought his way from a savage state to civilization, this should be reflected in our mythology and legends and in the folklore of the human race. This is especially true in view of the time factor since, according to the prehistorians, it was only some eight thousand years ago that the first settled agricultural communities came into being.

Written records of at least *some* kind are known to extend back to roughly 5000 B.C., so there is a gap of perhaps two thousand years between the first settled agricultural communities and the invention of writing and records. We could assume that village communities would have to devise some means of keeping records of their flocks and herds, particularly as regards ownership. This alone would be an incentive to creating some form of written language, perhaps a primitive picture writing. Tallies made on stone and bone showing phases of the moon, perhaps calendric computations, have been found which scientists date to an extremely ancient period, perhaps even tens of thousands of years ago. If people in the remote past did that, it is even more likely that they kept records of their possessions, and so eventually would produce written history.

In any case, several thousand years is not a great deal of time for oral history to survive. If one counts two generations to a century, information could be passed twenty times in a thousand years. We have evidence of tribal traditions and mythologies passed down for thousands of years. Many South American peoples and Polynesian islanders have remarkably detailed Flood legends passed down orally. These must be many thousands of years old, as this event must have taken place prior to 4000 B.C.

So records of man's emergence from a savage state to civilization, which one would think would be a matter of pride for any human group, should exist on a widespread scale. In fact, there are not even scattered references to such a state of affairs. Rather, the exact opposite is indicated. Most legends depict a state of advanced civilization, which was lost because of a great calamity for which the "gods" have taken a large share of the blame. The biblical fall of man is a case in point.

We are confronted with an intractable problem where humanity is concerned. If man had evolved from an apelike ancestor and then lived for tens of thousands of years as a primitive, what forces triggered the sudden emergence of highly advanced civilizations in the past? Further, it has been stated that the development of the human brain was so staggeringly rapid by evolutionary standards as to be almost instantaneous.

We have here a dilemma that the theoreticians are unable or

unwilling to admit exists. Regarding the past Ice Age it has been frequently stated that "no known natural forces which can be visualized can account for the glaciations or their sudden termination." With mankind we could say the same thing: no known *natural* forces could have shaped the human mind in such a short period of time, a period which can be measured not in hundreds of thousands but in mere tens of thousands of years. Such a natural force would be entirely contrary to the evolutionary concept of minute mutations building into major alterations over millions of years. Some biologists have thought that even if life as a whole has developed in an evolutionary manner, the period has been too short for the reasoning, human intelligence to have developed naturally.

Yet the evolutionary anthropologists and biologists cling doggedly to their theories—theories that are untenable because they do not fit either the evidence or the pattern. One must conclude that such scientists are unable to believe in the miraculous and are reluctant to admit to any other possibility.

In his book *Return to the Stars,* Erich von Däniken is aware of this problem and quotes Loren Eiseley as follows:

"Today on the other hand we must assume that man only emerged quite recently, because he appeared so explosively. We have every reason to believe that, without prejudice to the forces that must have shared in the training of the human brain, a stubborn and long-drawn-out battle for existence between several human groups could never have produced such high mental faculties as we find today among all peoples on the earth. Something else, some other educational factor, must have escaped the attention of the evolutionary theoreticians."

Von Däniken, from this, assumes that a *physical* evolution is possible (although more likely induced by deliberate and planned genetic mutation) but that the explosive growth of *intelligence* must have been artificially produced. His theory is that highly advanced aliens from another civilization in space, armed with a vast amount of biological knowledge, genetically altered existing human stocks to produce a high degree of intelligence.

These aliens biologically programed specimens for higher intelligence, he believes, at the same time implanting in this intelligence certain basic knowledge necessary for the development of civilization, including moral values and religious concepts. These religious concepts included awe and subservience to the "heavenly gods" (the aliens) so that the programing could be effectively carried out. Von Däniken states that the severe penalties instituted in many ancient communities for attempted mating with animals were instituted by the alien "gods" to prevent degeneration to a bestial state.

Such a speculation offers a solution of sorts to the sudden emergence of human intelligence, but there are many weaknesses in it.

For one thing, man cannot mate with animals, even those physically nearest to him—the anthropoid apes—*and produce offspring*. Each species carries within it a specific number of chromosomes; for example, the human being has twenty-three pairs and a bee sixteen. Only creatures with the same number of chromosomes can mate and produce offspring, which is why every separate species breeds true to type. The question of mankind producing bestial offspring simply *cannot arise*.

Even fossils of extinct apelike forms do not show any characteristics of *Homo sapiens*, which points to the fact that there could not have been any hybridization between human stocks and anthropoids or prehominid types. Modern man, *Homo sapiens*, who appeared on the terrestrial scene so abruptly with no apparent direct ancestors, is radically different both physically and mentally from any other anthropoid, past or present. To create him from existing stock, as Von Däniken postulates, would require the alien biologists to amend drastically both the brain and the physical structure.

It seems doubtful if any civilization from a planet elsewhere in the universe would go to such lengths to populate a different planet with intelligent life, particularly if they intended to create a species identical to their own. "After our likeness, in our image," says the Bible. This points to the fact that what our ancestors called the "gods" were human in form, although the exact relationship between man and the gods is an open question.

Hypothetical aliens, furthermore, would be unlikely to produce artificially an intelligent race which might one day become a menace to them. Von Däniken's hypothesis that certain knowledge was implanted genetically in the race, to be brought forth in stages, is a case in point. He suggests that the inventions of printing, the automobile, aircraft, etc., were neither accidents nor inspiration on the part of gifted men, but part of a "program" implanted by alien intelligences.

It is doubtful that a superior race would implant such dangerous ideas as space-flight capability coupled with the discovery of nuclear weapons and missiles. One would have to assume that if such aliens programed man to produce printing, medicine, and the like, they also gave rise to our fearsome arsenal of weaponry and the concepts of concentration camps, total war, and racial hatred—hardly what one would associate with an advanced and benevolent civilization!

It would be far simpler, and also much more likely, for any space-traveling civilization which discovered an earth-type planet not inhabited by intelligent life, to colonize such a planet by their own people. No doubt a civilization able to travel through space would have largely conquered illness and its own planetary environment. These achievements bring problems of over-population. Colonization of other planets may then become a necessity.

This has been our experience: the more we control our environment the more humans there are who survive, and this leads to overpopulation and the straining of all resources. Eventually, if our civilization survives and large-scale space travel becomes feasible, the stage could be reached when colonization of other worlds by us will become both desirable and necessary. Other civilizations may have encountered the same experience.

In our previous work, *Colony: Earth*, the possibility was raised that mankind may have originally been of extraterrestrial origin, the descendants of colonists from other worlds. This concept would explain the sudden appearance of humanity with all its knowledge and, further, would explain why there are three basic human groups: Negroid, Mongoloid, and Caucasian. Such theories as those of evolution, divine creation, and creation by artificial mutation fail to explain satisfactorily why there should be these differences. All human groups can interbreed, which shows that they belonged to a common stock *at some point in time.*

It is also true that the physical differences have no special bearing on survival capabilities in this particular planetary environment. Caucasian and Mongol occupy similar temperature zones, and the Negroid appears to have no difficulty in adapting to climates colder than those natural to him. We have mentioned in Chapter One an idea that man may have been a traveler in the galaxy for millions of years, moving from planet to planet as circumstances changed. Perhaps the original home of man is many, many times removed from this earth. Could it perhaps have been on the planets of other suns, with slightly different radiation emissions, that there was some adaptive protection in the differing makeup of these human groups, who had a common origin at some unimaginably distant point in time?

If we refer to our hypothetical Martian model in Chapter 1, we note that as we may one day send people of differing physical characteristics to another planet, other civilizations may have done so in the past.

Can we find any justification for advancing a case for extraterrestrial colonization? I think we can. Such an event should have left at least a trace in racial memory or in primordial folklore. The legends quoted by von Däniken, for example, to support his theories of gods from space could in fact better support the idea that it is man himself who was originally the god.

Another example is the curious Polynesian mythology, recorded by Bengt Danielsson, a companion of Thor Heyerdahl on the *Kon-Tiki* expedition. Danielsson noted a discourse of a priest called Te-Yho-e-te-Pange, on the island of Raroia in the Tuamotu group:

"In the beginning there was only empty space, neither darkness nor light, neither land nor sea, neither sun nor sky. Everything was a big silent void. Untold ages went by. Then the void began to move

1866834

and turned into Po. Everything was still dark, very dark, then Po itself began to revolve. New strange forces were at work. The night was transformed.

"The new matter was like sand, and sand became firm ground that grew upward. Lastly, the earth mother revealed herself and spread abroad and became a great country.

"There were plants, animals, and fish in the water, and they multiplied. The only thing that was lacking was man. Then Tangaroa created Tiki, who was the first ancestor."

Is this legend in fact a distorted description of a voyage through interstellar space until the solar system and then the earth is reached? A journey through space would be without day or night, land, sea, or sky. The first day would only commence when the planet's surface was reached and conditions that we consider normal were experienced. We observe in the legend that all plant and animal life was in existence in the world, *except man,* and this again agrees with our science, which states that man came last into the world.

Further, consider an American Indian epic called *Chon-oopa-sa,* ascribed to an Indian poet called Pa-la-ne-a-pa-pa and mentioned by Churchward in connection with his Mu theories:

> In the remotest past
> Millions and millions of moons ago
> The first of mortal men was cast down
> On this world by the great Wo-Kon.
> The first Dakota was formed from a star;
> He hurled him and watched him as he fell
> Through the darkness until he rested
> On soft soil. He was not wounded.
> Wa-kin-yan, the first Sioux.

Legends such as these are not unusual in North America.

The Canadian journal *Topside* reported: "The writer has recently met Chief Mezzaluna of the Piute tribe. In answer to the question, 'Where did the North American Indians come from?' the following was stated:

" 'According to our ancient traditions the Indians were created in the sky by Gitchie Manitou, the Great Spirit, who sent down here a big thunder bird to find a place for his children to live. He discovered this land . . . and brought Indians to settle on it. They were taught to use the land wisely and never abuse its natural resources.' "

On the other side of the world, the Soviet scholar Viaceslav Saitsev says in his book, *On Earth and Sea:*

"According to a Slavonic tale, 'Man was created far from the Earth and very long ago. When God had finished creating He commanded the angels to take some human couples to Earth so that they should multiply there. The angels spread the couples over the world, and wherever they set up home they multiplied. Perhaps when Earth is nearing its end, God will again take men somewhere else so that they may reproduce.' The mind which worked out such a tale must have been an elaborate one, fully developed. Though there may be fantasy here it is not without sense."

Clearly, we do have legends that connect man's origins with a "spatial" birth. The Polynesian description of space is scientifically accurate. The American Indian legend of the thunder bird is interesting, since it is widespread throughout North America and could have a basis in fact. A thing which flies and makes a sound like thunder—this reminds us of a rocket, which does make a thunderous noise. Eskimos also have a legend that long ago they were brought to their present homeland by great iron birds, although this may not necessarily mean space flight.

The Rigveda is the most ancient of the Indian sacred Sanskrit texts. From Paul Frischauer's book, *It Is Written* we quote the following:

"In those days there was neither not-being nor being. Neither the atmosphere nor the sky was above. What flew to and from where? In whose keeping? What was the unfathomable? In those times there was neither death nor immortality. There was not a sign of day and night. This *one* breathed according to its own law without currents of air. Everything but this was not present. In the beginning darkness was hidden in darkness. The life-powerful that was enclosed by the void, the one, was born by the might of its hot urgency. . . .

"Was there an above, was there a below? Who knows for sure, who can say whence they originated, whence this creation came?"

Again, an ancient prayer in the Egyptian Book of the Dead says:

O world egg, hear me.
I am Horus of millions of years;
I am lord and master of the throne.
Freed from evil, I traverse the ages
and spaces that are endless.

There is also the concept in ancient literature that the time standards of the gods from the sky were different from those of mortals of earth. For instance, a day of Brahma equals 4,320,000,000 years to a mortal. The Bible also says of God "that a thousand years in thy sight are but a moment."

What can we make of all these legends? Why should they exist? It may seem that they make very little sense unless we look at them within the context ot *spatial travel and Einsteinian physics.*

First, we have accurate descriptions of outer space: darkness, neither day nor night, no air. How could our primitive ancestors have known of these things? Without even aircraft, how could they have possibly known a point could be reached when there would be no atmosphere? Also, a state of neither day nor night would have been totally outside their experience.

Knowledge must have been passed down to them in some manner, and in spite of some distortions the descriptions are basically what long-distance space travelers would bring back with them. It is certain that without *experience* of these things, it would not have been possible to have written about them so accurately. The unimaginable, that which is *totally* outside the terrestrial frame of reference, cannot be imagined, much less with accuracy.

The Rigveda makes a curious statement when it says "there was neither death nor immortality." What can this mean? If it is not death or immortality, and presumably it does not mean normal mortal life, then it means something else, and this something could have been suspended animation. Translated into modern scientific terms, the statement in the Rigveda seems to refer to a journey in space: "Neither the atmosphere nor the sky was above." Were these space travelers in a state of suspended animation? "This one breathed according to its own law without currents of air." In a state of suspended animation, as envisaged by our scientists for long-duration space flights, neither air, food, nor drink would be taken. The travelers would be enclosed in an airtight capsule, with automatic life-support systems at a minimum setting. This is graphically described in the film *2001: A Space Odyssey.*

The reference to one who breathed by its own law could perhaps be referring to such a state; "its own law" as opposed to obeying the natural laws of life. "Living, but not breathing" could perhaps be a translation. Normal life would be resumed at the termination of the journey, when the earth was reached and the travelers were revived.

The reference in the Egyptian prayer seems to point to Horus being a space traveler. The curious statement that the gods who traveled in the voids of space lived at a different time rate from mortals is not so curious if it is taken in the context of space travel undertaken at near-relativistic velocities. At these speeds, the now well-known "time dilatation" effect is operative, whereby a journey lasting many light-years to people on earth actually takes a much shorter period for those undertaking the journey.

A round trip to Alpha Centauri, for example, would take approximately ten years at near-relativistic speeds (Alpha Centauri is 4.3 light-years distant from earth). To the travelers, however, the voyage would appear to have taken only a matter of weeks. This is not to say that time *itself* is altered or slowed, but that the apparent distance is decreased and therefore does not take as long.

This factor could lead us to question whether the space-traveling gods of antiquity in *actuality* were much longer-lived than earthbound people, or whether this was an effect of time dilatation. An astronaut traveling from a distant solar system could make visits to earth several times in his lifetime, whereas to the people on earth the visits would be at intervals of hundreds of years. The fact that many generations of earthbound people would be born and die in the lifetime of the visiting astronaut could well give rise to the impression of a visit by an immortal.

One could wonder whether the thousand-year visits of the god which is mentioned in the Hebrew and other mythologies is in fact connected with space voyages. If the interval between visits was a thousand years, it is possible an estimate could be made as to the actual distance traveled, and solar-type stars within a given radius from our system could then be suggested as the home star of the astronauts.

References in mythology frequently mention the seven stars of the Pleiades. This is actually a cluster of some hundreds of stars, but there are seven visible to the naked eye. These are sometimes said to be the home of the gods, and we have in the Jewish religion the seven-branched candlestick which has "heavenly" connotations. The Pleiades are situated 432 light-years distant, and it is perhaps significant that a round-trip voyage at slightly under relativistic velocities brings us close to a figure of a *thousand years.*

There is another curious legend from ancient India which may have a bearing on the methods employed in conveying colonists from one system to another. Traditions handed down to the Brahmins—the priestly class of India—say that lunar Pitris created life on this planet after their descent to earth from the moon. This would seem to suggest that the first men originated on the moon. The Brahmin tradition credited the moon as the cradle of life on earth and believed it was much older than the earth.

This seems somewhat odd, because in most legends the sun is thought of as the principal life-giving deity. It is also interesting to note the belief that the moon was older than the earth, for investigations resulting from the first lunar expeditions of the Apollo series have led some scientists to the conclusion that the moon's surface composition is so different from the earth's it may have originated outside the solar system, and also that it may be much older than the earth.

When the first Apollo astronauts left the moon they jettisoned the "lunar bug" after they had docked with the return capsule, and crashed it on the surface of the moon. The impact caused unexpectedly severe and long-lasting reverberations which immediately suggested that the moon was a hollow sphere, and as yet no alternative solution to such extraordinary echoes has been found acceptable.

The idea that the moon may be hollow is itself so extraordinary that it has been rejected by many scientists, since there seems no way in which a body, according to celestial mechanics, could have been formed hollow. Therefore it would have to have been accomplished by artificial means. The idea that the moon could have been altered by the action of intelligent beings is not one that would be accepted by our scientists.

It might be well, however, to consider just that possibility. The suggestion has been made that long-duration space flights could be undertaken without building huge spaceships which would have to be assembled while orbiting the earth or the moon. Their construction on the surface of the earth would give rise to insuperable problems in lifting them free of earth's gravitational pull. An alternative is to modify some of the larger asteroids which orbit between Mars and Jupiter and use these as vehicles. Such bodies, it is thought, could be hollowed out, fitted with life-support systems and propulsion units, and used as spacecraft.

The advantages are considerable. For one thing, the rocky shell could be of a thickness to give adequate protection against cosmic radiation and micrometeorites, and those which are of a roughly spherical shape would be ideal for long voyages in free-fall conditions. Spinning them as they traveled would provide an artificial gravity which, although less than earth's, would create more comfortable conditions for the travelers. An extremely long voyage in gravity-free conditions would cause difficulty in readjusting to a planetary gravitational field.

The difficulties in hollowing out such an asteroid and fitting it with support systems and propulsion units would be no greater than building a huge ship in earth orbit with the vast amount of materials which would have to be ferried from the surface. Admittedly, the distance to the asteroid belt is much greater than from a point in earth orbit, but by the time we are ready to initiate projects to ferry people to other solar systems the distance factor will have been greatly reduced by faster and more efficient propulsion systems.

This idea gives rise to a thought which is startling but neither impossible nor impractical: Did some other race in the past consider a similar solution, modify a large uninhabited body, and propel it through space from another planetary system—a body which later became our moon?

There are some peculiarities about the earth-moon relationship which have no counterpart in the rest of the solar system. No other planet except earth has such a large satellite, and it has properly been described as a twin planetary system rather than a planet and satellite system. Where the other planets of the system are concerned, the planet is invariably hundreds of times the size of any of its satellites, whereas the moon is a fifth the mass of the earth. Also, there are indications that at some time in the past the earth did not possess a moon, and it is thought that the moon is actually a "captured" body, caught in earth's gravitational field at some time in the past. Was the capture not accidental, however, but deliberate, with the moon parked in orbit at the end of its voyage, as we ourselves do with our artificial satellites?

A vehicle the size of the moon would be able to transport a large number of colonists, together with plants and animals which then could all be transported to the surface of the new world.

It may sound fantastic, but it does seem rather odd that some aspects of the Indian legend connect with modern discoveries about the moon. This hypothesis could also relate to the other origin myths about the long darkness, and the fact that the egg, or the sphere, is intimately connected with human origins.

Scientists were of the opinion that voyages to the moon would solve many problems that could not be solved by observational methods from earth alone. The results of the surveys have created more problems than they have solved, and the mysteries of the moon are perhaps greater now than before the astronauts went there.

Whether some of these puzzles will be solved by future astronauts only time will show, for numerous strange things have been reported by observers over the years. Many have claimed to see odd things on the moon through telescopes, and reports by trained and serious observers cannot be taken lightly. Moving lights have been observed, and domes that appear and disappear. There was a cross formation photographed by Robert E. Curtis, an astronomer of Alamogordo, published in the *Harvard University Review*. There is the strange block photographed by Sond 3 in July 1965 and given prominence in *Pravda*.

Then there is the matter of what is known as the Blair Cuspids. The attention of William Blair, a specialist in physical anthropology of the Boeing Institute of Biotechnology, was drawn to some photographs taken by Lunar Orbiter 2 of the western edge of the Sea of Tranquility and published by NASA on November 2, 1966. They are a group of monoliths on the lunar surface which cast very clear shadows, the tallest being more than 238 yards in height, and the others about the height of large spruce trees. These formations have attracted the attention of scientists on previous occasions but were dismissed as purely natural formations. (After all, people have thought they have seen buildings, bridges, and even canals on the moon, and these have proved to be natural features distorted by tricks of light and shade—and perhaps some imagination!)

About the Cuspids, Blair said: "If the Cuspids really were the result of some geophysical event, it would be natural to expect to see them distributed at random: as a result the triangulation would be scalene or irregular, whereas those concerning the lunar objects lead to a basilary system, with co-ordinates x, y, x, to the right angle, six isosceles triangles and two axes consisting of three points each."

Blair was asked if he considered this formation to have been the work of intelligent beings. "Do you want me to confirm it so that you can discredit me?" he replied. "Well, I will tell you this: if a similar thing had been found on earth, archaeology's first concern would have been to inspect the place and carry out trial excavations to assess the extent of the discovery."

Of course, this formation could be a natural formation of such regularity as to *appear* artificial. "But if this axiom had been applied to similar structures on earth," said Blair, "more than half the Maya and Aztec architecture known today would have still been buried under hills and depressions covered in trees and woods . . . 'a result of some geophysical event'; archaeology would never have developed and most of the facts relating to human evolution would have remained veiled in mystery."

Perhaps one day explorers will visit this particular part of the moon, investigate this mystery and prove whether the formation is natural or artificial. If natural, another phenomenon will have been solved; if artificial, we would have to wonder what purpose it served. Was it a navigational device, or a message? And for whom? Was it, if artificial, connected with the known mathematically aligned megaliths on earth? If so, this would point to visits to the moon at a period we call prehistoric. Only time will solve this and other enigmas of the moon.

However, whether or not the moon was originally a giant vehicle, spaceships traveling to earth could have been of completely artificial origin. It is a fact that the design of long-duration space vehicles now under serious consideration will be spherical or egg-shaped forms, as these are considered optimum for interstellar travel. In this way, of course, we are copying nature; most interstellar natural bodies are globular in shape. Freud has said that the sphere or the egg is one of the oldest archetypical images in the human subconscious and may in fact stem from an ancestral memory of such vehicles. The golden egg which descended from the sky is a theme of some mythologies, particularly those of the Pacific and Easter Island.

We also have to consider a common human experience;

dreams of floating or flying, which astronauts have likened to their experience of weightlessness. This sensation is so exhilarating that astronauts have frequently to be warned about their reluctance to terminate their "space walks."

Certain areas of experience, stemming from the deep unconscious and revealed only in the dream state, are reckoned to be ancestral memories, inaccessible in the conscious state. One explanation for the floating sensation, so akin to levitation, suggests that it stems from the time when all life lived in the sea, where the effect of the earth's gravitational pull is not so noticeable. However, this subconscious experience does seem to have more in common with the astronauts' experience and may in fact relate more to an ancestral memory of the time when human beings traveled in space and experienced a state of weightlessness.

Descriptions of the vacuum of space; dreams of weightlessness; golden eggs from the Gods in the Sky—it is curious to note here that the living quarters of the lunar landing vehicles were coated with gold foil, a reflective agent to prevent excessive heating. Do not these things point to a possibility that in the past earthbound man was cosmic man?

3

Fall of Angels

THE FLOOD is a worldwide legend; the condition of the world and of humanity *before* the Flood is also the subject of much mythology, and these legends from different points of the globe have many points of convergence. It is generally recognized that the Flood legends in the New World did not stem from the Sumerian/ Babylonian myths, and therefore it seems possible that a planetwide catastrophe occurred which was not related to the localized floods of the Tigris-Euphrates delta. (This aspect was fully discussed in *Colony: Earth.*) From this premise we can postulate that the state of the world before the Flood, independently described by many different groups from all over the world, must have had a basis in reality.

One common theme is that of a vanished Golden Age which perished with the Flood. Also widespread is the idea that before the Flood man had access to a great deal of knowledge which made the gods fearful, so they caused the destruction of the majority of the human race, and the loss of all knowledge.

There are accordingly two main themes behind the reason for the apparently divine visitation of the Flood: man's wickedness, which is stressed in the Old Testament, and his acquisition of great knowledge, which is described as "sin" in the Bible.

We must seek elsewhere in mythology for the knowledge

referred to. For example, members of an Indian tribe in South America say in their legends that men learned how to fly and so the gods destroyed them. In the Maya *Popol Vuh* it is said that the "first men" could see what was far away and what was very small, and they surveyed the four quarters of the globe, but then the gods closed the eyes of the first men, and all their knowledge was lost.

It is logical to assume that if an advanced civilization was largely destroyed during a planetwide catastrophe, most of their knowledge *would* be lost as the survivors reverted to savagery.

We can infer the existence of such an advanced civilization, which was probably worldwide in extent, partly from legend and partly from certain evidence, both material and documentary. First we shall deal with the legendary evidence.

The myth of Paradise before the Flood is common to all the ancient Middle Eastern cultures, the most familiar, of course, being the biblical Garden of Eden where there was no sickness or knowledge of sin. The Sumerian legend is almost identical to the biblical description; the Sumerian poem quoted here is called by Dr. Kramer the Epic of Emmerkar:

The Land Dilmun is a pure place, the Land Dilmun is a clean place.
The Land Dilmun is a clean place, the Land Dilmun is a bright
   place.
In Dilmun the raven uttered no cry.
The kite uttered not the cry of the kite.
The lion killed not.
The wolf snatched not the lamb.
Unknown was the kid-killing dog.
Unknown was the grain-devouring boar.
The sick eyed says not, "I am sick eyed."
The sick headed says not, "I am sick headed."
Its [Dilmun's] old woman says not, "I am an old woman."
Its old man says not, "I am an old man."

In the Semitic version of this myth Dilmun was the dwelling place of the immortals.

We note this close similarity between the Sumerian and Hebrew myths: the absence of sickness and the lack of predators, so that the domesticated animals were always safe. It may not be important to argue whether the Hebrew was a copy of the Sumerian myth, or whether they developed independently. What is important is that all these myths speak of a *condition* which existed. If this had been mere wish-fulfillment, surely the ancient chroniclers would have said: "One day there will be no sickness, and the lion will not kill the lamb, etc." They all appear to be quite convinced that this state of affairs existed at a period before the Flood. Why, then, should we automatically assume that this was merely a story?

We can, I think, make a parallel between our own and biblical times. If we could transport a dweller from the Palestinian desert region of about 1000 B.C. to present-day England, what would his impressions be, and, more important, what impressions would he take back with him when he was returned to his own time?

He would find a countryside so thoroughly cultivated it had the appearance of a garden. He would find the cattle and sheep perfectly safe in their fields, so safe that they could be left unattended all day; no wolves, lions, bears, or eagles here. The people also would be so free of diseases as to seem miraculously healthy; no trachoma or leprosy, plague or cholera, those common scourges of primitive times.

Would he not, in fact, be seeing the legendary Paradise described to us by ancient authors?

Of course, we know our world is no paradise, but it might seem so in retrospect to our descendants if they were to revert to savagery and the world descend to wild and uncultivated nature. So it seems possible that an advanced civilization may have enjoyed conditions in the remote past similar to those prevailing today.

The people of this civilization would have been as free from disease as we are today. We are not, of course, disease-free; we have cancer, heart ailments, etc. But the principal killer diseases of the past have been virtually eliminated from civilized countries. There is no leprosy or cholera; tuberculosis is almost conquered; bubonic plague and typhus are rare events and swiftly dealt with.

The basic problem we have to face is: how did such an advanced culture come into being at such a remote period?

If humanity had colonized this planet from another world in space, then they would have brought, if not the apparatus, the *knowledge* of that civilization with them. Perhaps these remote ancestors of ours from the stars came here already completely disease-free and also with greatly extended life spans.

There is a consistent theme running through all legendary sources regarding the people from before the Flood. This is that they were descended from the gods, and in many instances were a hybrid of man and god.

Quoted here is an excerpt from *Middle Eastern Mythology* by S. H. Hooke (Penguin, 1963), page 132:

"The myth of the union between divine and mortal beings, resulting in the birth of demi-gods or heroes, is found in the early Sumerian and Babylonian sources whose influence on Canaanite mythology appears in the Ugaritic texts.

"Behind the brief and probably intentionally obscure reference in 6:1–4 there lies a more widely known myth of a race of semi-devine beings who rebelled against the gods and were cast down into the underworld. The beings, called Nephilim in verse 4 (Genesis), and rendered giants in the Septuagint and Authorized Version, seem to have been regarded by the Yahwist as the offspring of the union between the 'Sons of God' and the daughters of men mentioned in verse 1. The assembly of lesser gods, so often referred to in Sumerian, Babylonian and Ugaritic myths, have been transformed in Hebrew myth and poetry into the 'Sons of God' conceived of as a kind of heavenly council over which Yahweh presided. Compare for instance, the scene in the first Chapter of Job, where the Sons of God come to present themselves before Yahweh (Job 1:6). Traces of the myth are to be found in Num. 13:33, where the Nephilim are represented as the survivors of a race of giants whom the Hebrews found in Canaan when they came to settle there. Another possible reference occurs in Ezek. 32:27, where a slight emendation gives us an allusion to the Nephilim. In apocalyptic literature and in the New Testament (2 Pet. 2:4: Jude 6) the myth has been still further transformed into the myth of the fall of the angels, so splendidly portrayed by Milton. The fragment of the myth here preserved by the Yahwist was originally an etiological myth explaining the belief in the existence of a vanished race of giants, but the Yahwist has made use of it here to support his account of the progressive deterioration of the human race, and goes on to connect it with Yahweh's purpose to destroy Man from the face of the Earth.''

There are three principal threads running through this mythological series: one, the existence of a race of semidivine beings; two, the casting down from heaven of this race of beings (sometimes regarded as giants) and the instance of their being "chained in the earth"; and three, the progressive deterioration of the human race.

How did ancient peoples happen to devise such a complex mythology? A mythology of a race of supermen, destroyed in a Flood; the destruction of most of the human race, and the gradual degeneration of the survivors. Surely these myths must be based on factual events, no matter how distorted.

If we were to assume that mankind had originated from spaceborne colonists, as suggested by some of the myths of origin, then these colonists could have become, in the course of time, associated with the gods who came from the sky. These earlier-generation "space gods" would, once settled on earth, produce children. Would this perhaps be the union between the gods and terrestrial man? Later generations would be regarded as the product of gods and men, and thus semidivine.

We suggested in *Colony: Earth* that an interstellar traveling society may, by its very nature, be composed of beings of (to us) greatly extended life spans. As a civilization develops its sciences, one of its aims would be the conquest of disease and the aging process. Hand in hand with the advancement of the medical sciences would go the progress of technology. Long-distance space travel may well accompany progress in medical matters. Lengthy space voyages would more easily be accomplished by creatures with greatly extended life spans. A voyage lasting thirty years is half the life of a present-day terrestrial human; to a being with an average life span of five hundred years such a voyage would be relatively short, and of little consequence as far as the aging of the astronaut is concerned.

It is possible that by the time we are ready to undertake long-duration interstellar flights in manned vehicles, the problem of aging may largely have been solved and human lifetimes dramatically extended.

It is perhaps pertinent to point out that the Old Testament tells us that the generations before the Flood possessed greatly extended life spans. Although the ages mentioned in Genesis of some individuals perhaps cannot be taken too literally (Adam was reckoned to have lived for 930 years, Seth 912 years), it does point to a belief that the generations before the Flood lived a very long time. Such a belief may well have been based on fact. We cannot deny that a superior civilization may have conquered the aging process; if such a thing were impossible we would not be attempting at this time to discover the cause of aging and arrest it.

Turning to another aspect of the ancient myths, consider the belief in the existence of giants which is widespeard in many mythologies, occurring in the Middle East and among the Aztec and Maya mythologies of the New World. One explanation of this may stem from the many odd ruins which the people who devised the myth will have seen: ruins built on a cyclopean scale, constructed of blocks weighing scores and even hundreds of tons. The people of 2000 or 3000 B.C. were quite incapable of raising such structures and may well have thought they had been built by giants.

One may consider the Pyramids of Egypt, great artificial mountains constructed of blocks weighing up to ten tons and with beams weighing more than fifty tons; the huge temples of Malta with blocks weighing hundreds of tons, and the terrace of Baalbek in Syria; also constructed of massive blocks. The vast scale of the pyramid complex at Teotehuacan in Mexico inspired awe in the Toltecs, the forerunners of the Aztecs, and huge-scale construction is common in the ruins of the Inca Empire.

Similarly, in Europe legends exist of giants, and here too there are great structures composed of massive stones, such as Stonehenge, the chambered tombs, and the dolmens. Right up to late medieval times it was commonly believed that these had been built by magic (and therefore, the early church thought, by demons). People lacking any kind of technology would well think such structures to have been erected by giants, as only they would have the strength to move such stones. It is perhaps significant that many of the legends regarding giants stem from areas where such ruins exist, and in many cases are associated with them.

Whether such giants actually existed is doubtful, although skeletons are supposed to have been discovered of some. However, the discoverer invariably says that the bones crumbled to dust once exposed to the light of day, which seems somewhat odd, as other bones, including those of long-extinct dinosaurs, have survived this event.

It is possible that a race of people existed in the past taller than the majority of present-day peoples, as skeletons of what are regarded as early true *Homo sapiens,* called Cro-Magnon, are on average some six inches taller than we are, and with a larger cranial capacity. However, such people could scarcely be regarded as giants, and it is more likely that it is the technology of an earlier race, rather than their physical characteristics, which has given rise to the legends.

Pursuing these myths of a Lost Eden further, can we formulate a hypothesis to explain the rebellion of the semidivine beings (sometimes called Nephilim) who were cast into the underworld? If we start with the premise that there existed in the remote past a highly advanced civilization, possibly extraterrestrial in origin, we can postulate that the legend is actually a gross distortion of real events.

We have to consider the basic tenet that the semidivine beings were actually the descendants of a space-traveling civilization, and this means that there existed elsewhere the civilization that sent them here. If this planet had been selected for colonization by a Superior Community, that group would be aware of the existence of a culture developing here. Further, we have to consider the possibility that when the new civilization became established, they developed (or redeveloped) space travel, and contacts and communication existed between them and the Superior Community elsewhere in the galaxy.

Was the destruction of the human race in the Flood and the casting down of the semidivine beings the result of a conflict between these two civilizations? This may explain the legends of the War in Heaven, the War of the Gods, which is related to the episode we have just mentioned.

This possibility was dealt with in *Colony: Earth,* and we may just mention here that certain passages in the Old Testament can be brought together and related to such an episode.

There are several curious statements in the Old Testament, particularly in Isaiah, which have never been properly understood.

Consider, for example, this:

13:4, 5: "Hark, a tumult on the mountains as of a great multitude.

"Hark, an uproar of kingdoms, of nations gathering together.

"The Lord of Hosts is mustering a host for battle. They come from a distant land, from the end of the heavens, the Lord and the weapons of his indignation, to destroy the whole earth."

Also, Isaiah 13:13:

"Therefore I will shake the heavens, and the earth shall move out of her place, in the wrath of the Lord of Hosts, and in the day of his fierce anger."

These two passages would seem not to be related to any localized events, but to connect more closely with the Flood destruction in the Genesis narrative. Isaiah also says that the earth shall be laid waste, and men become more rare than fine gold.

The description of a Lord who came with unnamed fearful weapons and a host from a *country at the end of heaven* has never been given a satisfactory explanation. Translated into modern terminology, does it not sound more like an arrival from another world in space? A country at the other end of heaven—thus may ancient peoples unfamiliar with astronomical truths have described another world in space.

The passage that says the earth shall be moved out of her place has often been considered to be a fanciful description of an earthquake, but this may not be so, since earthquakes are mentioned freely in the Old Testament, and in more readily understood terms. There is a great deal of evidence to show that a great catastrophe in the past (as opposed to the Ice Age theory) could have been caused by a disturbance to the earth's orbit around the sun. This point was covered in *Colony: Earth* and need not be repeated here.

Much of the Old Testament is a record of catastrophes. These catastrophes consist of three main events: the Flood in Genesis; the Plagues of Exodus, which are connected with the great volcanic eruption on Thera (Santorini) in 1500 B.C., and a great destruction *which is to come.*

Most of the world's mythology devotes a great deal of space to *two great disasters:* the Flood and a great disaster at some time in the future. Vedic, Tibetan, Maya, Inca, and Aztec mythologies were much concerned with the future destruction, and the earth's history was divided into World Ages, or cycles of disasters.

How did they arrive at such conclusions, and why did they all consider there was a great disaster to come which would utterly change the face of the earth?

It is possible to formulate a hypothesis regarding these legends, but only if we assume the occurrence of a planetwide catastrophe in the past connected with a cosmic event, either natural or artificially engineered.

The destruction of a highly advanced civilization in the past by a global disaster would still leave survivors—it obviously has, or we would not be here today. Some of these survivors, the scientists and mathematicians, would wish to estimate the extent of the disaster which had overtaken the planet. To this end they may have devised the mathematical and astronomical complex of which Stonehenge is part.

Their calculations would show that the earth had been disturbed in its orbit, and that this disturbance, by moving the planet into an *outward* spiraling path, would lead to disastrous conditions for survival of life in some thousands of years. The world would most likely very slowly become colder, until the time came when survival would become difficult. This minute movement of the earth, operating now for six or eight thousand years, has led to a slow and steady deterioration in the world's climate, and it is possible that this could worsen during the next several thousand years.

This is not mere fantasy or idle speculation, as it is obvious that the climate is deteriorating. Many areas of the world which once harbored prosperous cities are now deserts, and many of the deserts are growing larger. The planet is steadily cooling, as evidenced by tests that revealed the ocean temperatures are lower than they used to be. The Antarctic ice cap is enlarging at the rate of many millions of tons of ice a year. Far from emerging from a hypothetical Ice Age at the termination of the Pleistocene, we seem to be steadily moving into colder conditions. It has been thought recently by meteorologists that one possible cause could be that the earth is *moving slightly away from the sun,* although they do not suggest a prior cause as to why this should be happening.

We say it is this factor, this ancient *mathematical and astronomical knowledge,* which is the truth behind the Armageddon concept. The knowledge that the earth had been started on a course to disaster has, over the ages, been transformed into a religious-orientated prophecy of divine vengeance and punishment.

We observe that the Old Testament and Revelation are concerned with three disasters, while mythologies from other parts of the world mention *two.* This points to the one—the Plagues of Exodus—being a localized phenomenon, and the other two being planetary events.

Not only Genesis but many of the prophets are much concerned with the destruction of mankind by these catastrophes, and this is true also of Revelation. It would appear from this biblical summation, if one may use the term, that this is a compound of the Flood legend, the Thera eruption, and Armageddon (the disaster to come).

We note in Revelation 12:7: "Now war arose in heaven, Michael and his angels fighting against the dragon and his angels, and the dragon and his angels fought, but they were defeated and there was no longer any place for them in heaven. And the great dragon was thrown down, that ancient serpent, who is called the Devil and Satan, the deceiver of the whole world—he was thrown down to the earth, and his angels were thrown down with him."

Revelation 20: ". . . and he seized the dragon, that ancient serpent, who is the Devil and Satan, and bound him for a thousand years, and threw him into the pit, and shut it and sealed it over him, that he should deceive the nations no more, until the thousand years were ended."

It is necessary here to relate this to the statements contained in Genesis. We remember in our quote from *Middle Eastern Mythology* that the Nephilim, semidivine beings who were God's servants and dwelt with him in heaven, had *rebelled and were cast down*. Therefore the Nephilim and Satan and his (fallen) angels *are one and the same thing*. As the casting down of the Nephilim took place during the *Flood disaster*, then this passage in Revelation refers obviously to Flood events.

Therefore, the War in Heaven between God and his angels and Satan and his angels is part of the Flood episode. It shows—supported by mythologies from elsewhere in the world—that the Flood itself was an effect of a vast planetary disaster.

We have postulated that this catastrophe was not natural but more probably artificial, involving weapons of appallingly destructive powers, and there exist descriptions of weapons that appear to resemble nuclear bombs. Even the description of radiation sickness is clinically accurate (see the *Mahabharata*).

This possibility of a nuclear holocaust in the distant past raises some interesting points when attempting to formulate reasons for statements in Revelation, particularly those dealing with the fallen angels and Satan's being cast down and sealed under the ground for a long period of time. Biblical numbers appear to have been ritualized—the forty days of rain of the Flood, Christ's forty days in the wilderness, the thousand-year visits of God, and the casting down of Satan and his angels for a thousand years. One suspects that the one stands for a short period of time and the other a long period of time, and that they are not to be taken too literally.

We postulate that the War in Heaven was an actual war, possibly using nuclear weapons. In a nuclear conflict one protection from radiation effects would be shelters deep underground—this is the sort of precaution we ourselves have provided against a nuclear disaster. In such a war in the remote past it is logical that such shelters would have been built, not to protect the mass of the people, but to shield the scientists and the leading members of the government against the day when it would be safe to emerge and commence the task of rebuilding civilization. The passage in Revelation 20 could in fact refer not to a mythological event but to a real one—the casting down was, in fact, hibernation in deep shelters against radioactive contamination.

As we are all aware, radiation effects can be lingering. Radioactive cobalt, for example, has an extremely long half-life, and a cobalt bomb could render its point of impact uninhabitable for fifty thousand years. It may have been necessary for those "entombed" in these shelters to stay there for many years, perhaps even several generations, until the radioactivity level had dropped to tolerable limits. Can we not suggest, therefore, that the sealing of Satan and his angels in a pit for a thousand years was in fact the subterranean incarceration of important sections of that civilization for many years?

There is startling verification of such an idea. What would best support that hypothesis would be the discovery of such deep underground shelters. We have already suggested that the so-called passage graves were in fact hastily constructed shelters against either attack or the devastation resulting from the catastrophe. It has now been shown that such deep, artificially constructed shelters do in fact exist.

Erich von Däniken, in his book *Gold of the Gods* (Souvenir Press, 1973), has revealed the existence of a network of tunnels and caverns, at least partially artificial or artificially enlarged, almost a thousand feet below the surface in Ecuador, South America. The fact that these deep underground chambers were built and occupied by man has been proved by the artifacts found in them. In a huge chamber the size of a modern aircraft hangar were a large table and chairs made of an as yet unknown material. Golden statuettes and some three thousand leaf-thin gold metal plates inscribed in an unknown language also were found.

The inscriptions on the plates, which seem to be more alphabetical than hieroglyphic, would appear to bear a resemblance to ancient Cretan Linear script and to Sanskrit. This in itself could point to an extremely ancient written alphabetical language, completely unknown to us, which has undergone great transformations in the course of time, becoming the alphabetical languages of the ancient world, although greatly modified.

Underground chambers and tunnels, built on a megalithic scale, have been found in Turkey (the site of man's most ancient identifiable settlements), and underground passages exist beneath the cyclopean structures at Sacsayhuaman above Cuzco, in Peru.

Von Däniken's explanation for these extraordinary underground chambers is that they were built by spacemen who visited this world thousands of years ago, as presumably any people indigenous to this planet were incapable of such feats.

But why should it be assumed that everything extraordinary was constructed by visitors from another world? It has even been suggested that the pyramids and Stonehenge were built by aliens from elsewhere, because these structures do not fit with the picture we are given of the capabilities of our remote ancestors.

Weighing the mass of other evidence—the Flood legends, the "age of the gods," and the suspicion that there existed a superior civilization in the distant past—does it not seem more likely that those places were built by terrestrial humans in an attempt to ensure the survival of at least some of the human race against the dangers of a vast disaster? What possible reason could spacemen have had for building such places? If danger threatened them, all they would have to do is to lift their ships into space.

It is these places which could have been the *pit of Revelation*. Possibly many more such shelters await discovery in many parts of the world.

The tunnels in Ecuador were first discovered by an Argentinian scholar, Juan Moricz, following clues supplied by Ecuadorian Indian folklore. Similar legends exist among the Indians in the Maya region of Central America, which tell of underground cities and tunnels stretching for miles beneath the ruins of the ancient Maya centers. In the light of this new evidence, we should not be too hasty in discounting these legends.

Does the discovery of these once inhabited underground chambers also throw a new light on the widespread legends of the underworld, where dwelt the "gods of darkness?" We have suggested that the "gods of the sky" were earthmen traveling in aircraft and possibly also astronauts from elsewhere in the universe. The underworld gods, by the same reasoning, were those members of a superior civilization who had been forced to live underground for great periods of time, while their less fortunate contemporaries had to endure the hostile conditions of the surface, bereft of civilization, reverting to barbarism and distorting truth into legends.

The Sumerians had their legends of the underworld, where they described the journey of Inanna, Queen of Heaven, to her sister Ereshkigal, goddess of the underworld.

The Babylonian form of this legend is that of Tammuz and Ishtar.

In Ugaritic mythology Baal visits the underworld.

The Egyptians had their legends of the underworld, as did the Maya, the Celts, and the Nordics. In fact, there is scarcely a place on earth where this myth does not exist in some form or other.

It is thus possible that these legends were based on fact, although as in most mythology the facts were distorted and incorporated within the general framework of religious myth and ritual.

We are brought to a further point, and that is the emergence of that part of religious tradition regarding those mysterious figures, the "culture gods," or "culture bearers." Every legend of the building up of an ancient civilization says it has always been accomplished with the help of these godlike figures who guided their savage followers to the paths of culture. They were the ones who taught building and metalwork, agriculture and irrigation, mathematics, astronomy, and law.

Osiris of the Egyptians, for example, was a lawgiver and the teacher of healing.

Wirakocka of the Peruvians taught the arts of civilization, plant and animal husbandry, building and irrigation.

Kukulkan of the Maya and Quetzalcoatl of the Toltec/Aztec taught all the arts of civilization.

Each of the Greek gods had a specialty he taught to mankind.

The Babylonians had their legend of the Apkullu, who had taught their ancestors the rudiments of civilization.

Whence came these strange superior beings?

Some, it was said, came from the sky. Others apparently came from nowhere. If they had emerged from their long incarceration in *secret chambers deep underground, it could have been thought that they had emerged from nowhere.*

The gods from the sky and from the darkness deep underground—perhaps they both played their part in leading their barbarous descendants once more into the light.

4

World in Decline

THE MYTH of the degeneration of man, recounted in Genesis as the fall following the Flood, is also a worldwide phenomenon. The wording may be different; the general impression given is the same. Such myths are common, as may be expected, throughout the Middle Eastern and Mediterranean civilizations, where possibly one has influenced the growth of the other, but they also exist in one form or another in many other parts of the world.

In the sacred book of the Maya, the *Popol Vuh*, it is said:

"They looked into the distance and could discern everything which was in the world. When they looked they saw everything around and the dome of the sky and the inside of the earth. Without moving themselves they saw everything hidden in the distance. They saw at once the entire world, from the place in which they stood. Their wisdom was great. Their eye reached every forest and mountain and lake, every hill, sea and valley. Verily, they were wondrous men.

"The Gods said: Let us satisfy their desires a little, for what we see is not well. Must they resemble us in the end, their creators who know and see all from the distance?

"So the gods cast a veil over their eyes so that they grew dim as when the breath touches a mirror; they could see only that which was near at hand and clear. Thus was destroyed all the wisdom and knowledge of the first men."

These "first men," also called the Saiyam Uinicob (Old White Fathers), were reputed to have built the first cities and to have been destroyed in the Flood.

South American Indian tribes have the legend of the time when men had learned how to fly, and the gods became afraid that they would rival them and destoyed them.

Constantly there is this theme of man's former greatness, his near-destruction, and the reduction of the survivors to ignorance, deprivation, and disease.

The thing that concerns us now is, Can we establish from the ancient myths a basis for belief that a superior humanity existed in the past, and that this humanity has degenerated and may even still be degenerating?

There is a certain amount of evidence that could point to a degeneration of the human species, and we could draw conclusions from some aspects of the present human condition to support such a hypothesis.

We are all aware of the great longevity attributed to men in the period before the Flood, as mentioned in Genesis. Such legends also exist in Sumerian and Egyptian mythologies. We are also aware that there is, in Genesis, mention of a systematic reduction in the life spans of the generations following the Flood. We suggest that this would be a logical sequence if humanity were on the road to degeneration. Each generation would be less healthy than the one which preceded it, and a decrease in life span would be a natural consequence.

The present life span is some seventy years, in spite of all our advances in hygiene and medical techniques, which is exactly that mentioned in the Bible for the then "present" generations of man. In the less developed societies, such as India and Africa, the average life span is *half* of this. Actually, what we in the West have achieved is not a prolongation of life, but, by improving living standards and medical care, we have ensured that a majority of our citizens survive the hazards of existence to reach this "biblical" age of seventy. Actual extension of the life span has not been achieved as yet.

We have already mentioned the advantages of an extended life span for a space-traveling community, and colonists to earth from such a community could have been long-lived, which have given rise to the legends about the demigods from the Golden Age.

It may well be that there are areas of evidence for supposing that man may once have been a longer-lived species than at the present time.

First, let us consider that remarkable mechanism, the human brain.

It has been estimated that there are some ten billion gray cells, or neurons, in the brain. Each neuron contains upwards of 20 million RNA molecules (RNA molecules are the DNA "messenger" molecules) and each RNA molecule is capable of handling millions of "bits" of information. It has been estimated that during a normal human lifetime the human brain absorbs some one billion billion "bits" of information. As Isaac Asimov, the American science writer and biochemist, has said:

"There is no question, then, that RNA presents a filing system perfectly capable of handling any load of learning and memory which the human being is likely to put upon it—and a billion times more than that quantity too." (*New York Times Magazine,* October 9, 1966)

The average human being of today utilizes only the minutest fraction of the brain's potential, and it would appear that there are large areas of the brain which lie dormant and have no known function. Yet possibly everything we experience and learn in a lifetime, down to the most minute and insignificant detail, is stored away permanently somewhere within this vast organic memory storage unit. It is known that under hypnosis, or sometimes in dreams, information can be brought to the surface from the subconscious which cannot be reached by the conscious mind, or by a conscious act of will in the normal way.

Occasionally we hear of people who possess remarkable

memories, or of those who can handle enormous numbers and compute huge and complex sums in their heads almost instantaneously—problems that would take most people weeks to work out on paper if they could do them at all. There is the example of Shakuntal Devi of India, known as the Human Calculator, who was asked by the University of New South Wales in Australia to appear in a contest against UTECOM, one of Australia's biggest computers. So in Sydney, in 1959, Miss Devi was asked: "Find the cube root of 697,628,098,909." She gave the answer in seven seconds: 8,869. The computer provided a slightly different result, which was wrong, as a rerun proved.

One could say that an electronic computer is a large-scale version, or imitation, of the human brain, which contains only a tiny fraction of the relays contained within the brain. A computer with several million relays is capable of producing calculations in a matter of seconds which would take teams of mathematicians months to compute using paper and pencil. Yet these computers can handle only a millionth of the information the human brain can absorb. The difference is that any fact stored in the memory section of the computer can instantly be recalled, whereas in the case of the human being the greater part of the memory circuits are locked away in the subconscious and are largely inaccessible.

If the human brain functioned as efficiently as an electronic computer, then a human being would be able to recall every detail of everything he has ever learned and, furthermore, would be able to analyze and compute from these data. As we have seen, there are a few people who, by a "freak" of nature, are able to do this on a limited scale. But are they really the freaks, or is it the rest of us who are?

When one considers the way the brain functions, or rather should function, and compares it to a computer, which is an electronic version of the brain, then we see that if our potential were realized we would all be able to act in this way. How the few gifted brains function is not at all understood. The answer seems to lie in the way the recall system operates, and this would appear to function at a low level of efficiency in "normal" people.

If full use of the potential of the brain was realized, then we would remember everything we had learned. No subject would need to be learned painfully and slowly over the years, since a fact once known would never be forgotten.

But even if the potential of the brain were realized, it would still be capable of absorbing vastly more information than could be gathered in a present normal life span. Possibly a man could live for a thousand years and still not reach the full potential of the brain's information intake. It would appear on balance, then, that the brain was designed for a creature with a vastly longer life span.

If we consider that the human brain is not utilized to its full extent, does this point to an evolving organ? According to the evolutionary theory, animals evolve from simple to more complex forms, and this should hold true of human beings. These changes are reckoned to take place over periods of millions of years, and yet *Homo sapiens* appears to have "evolved" explosively over a period of less than half a million years. The "explosion of mind," as it is termed by the anthropologists, was accomplished so rapidly (by geological reckoning) as to be almost instantaneous. By this same reckoning, the human brain should still be developing, and there should be noticed, over the last six or seven thousand years, an appreciable increase in individual human intelligence. There is no evidence to support such a view—rather, the reverse is true.

The brain size of modern man averages out between 1,350 and 1,500 cubic centimeters. Our first true ancestors, Cro-Magnon, who lived some thirty to thirty-five thousand years ago, had a larger brain than modern man. The so-called Old Man of Cro-Magnon had a brain capacity of some 1,600 cubic centimeters. *Time* magazine of March 19, 1961, gave the brain capacity of Neanderthal man as 1,625 cubic centimeters, considerably greater than present-day humans.

On average, then, both Cro-Magnon and Neanderthal man had a brain as large as modern man's, or larger.

Cro-Magnon was first discovered by a French archaeologist, Louis Lartet, in a rock shelter in a cliff face near the village of Les Eyzies in the Dordogne in 1868. The shelter was called Cro-

Magnon, hence the name assigned to the skeletons of the people discovered there. Analysis of the skeletons showed them to belong to people with long straight limbs, with an average height of more than six feet. The forehead was exceptionally high and domed, and the brow ridges were so reduced as to be almost indiscernible.

We now turn our attention to Neanderthal man, whose origin is as obscure as modern man's, and about whose pedigree there is considerable discord in anthropological circles.

Homo neanderthalensis is named after a seventeenth-century German theologian called Joachim Neander. In honor of his ability as a writer of hymns the citizens of Düsseldorf named a valley near their city the Neanderthal, and in this valley the first bones of Neanderthal man were discovered. The discovery was made by workmen cleaning out a limestone cave, and most of the skeleton was thrown away by them—the surviving parts being a skullcap, some limb bones, a few ribs, and part of a pelvis. A Dr. Fuhlrott of Eberfeld collected these specimens, and Professor D. Schaaffhausen of Bonn gave the first scientific description of them.

This Neanderthal has been pictured as follows: erect-walking, stocky, and barrel-chested, with his head sunk somewhat between exceedingly broad shoulders, and with short legs and powerful arms. The head had a receding brow and a chin which seems too large for the body.

However, many kinds of Neanderthal man have been found, and they appear to have occupied as wide an area as Cro-Magnon or modern man, being found all through Europe and the Middle East. The Ehringsdorf and Steinheim skulls from Germany and those from Mount Carmel and other places in Israel appear to be more modern in appearance. The Ehringsdorf skull possessed heavy brow ridges but was high-domed, as are the Cro-Magnon skulls.

Briefly, the anthropologists divide Neanderthal into two classes: the "conservative," rather like the traditional picture of an apeman, and the "progressive," which seems to be a combination of "ape type" and Cro-Magnon. It has also been found that Neanderthal and Cro-Magnon (or *Homo sapiens*) existed at the same time. There may have been a degree of interbreeding between the two types, with, in the end, *Home sapiens*, represented by Cro-Magnon, absorbing or eliminating Neanderthal.

If it were true that the two types could interbreed, it must prove that they were from a common stock, and not from divergent lines of development; otherwise fruitful unions would not have arisen. It is a little difficult to imagine that the advanced-looking Cro-Magnon, who had, apparently, a more "intellectual" appearance than even we ourselves, would mate with the Neanderthal of the "conservative" type as he has been pictured by the anthropologists—a shambling, hairy brute differing little in appearance from the apemen reconstructions we have been shown.

The age of Neanderthal man is still a very open question, but in view of the possibility of both conflict and hybridization with Cro-Magnon, he falls within the chronological period of true humanity. It is generally assumed that he lived between an upper limit of 160,000 years ago and a lower limit of 35,000 to 40,000 years ago, being supplanted thereafter by *Homo sapiens*.

Neanderthal man, in spite of his apelike appearance, nevertheless had important human characteristics. He was familiar with fire and used stone tools, although it is the opinion of some experts that some of the stone flakes which have been taken to be his tools are actually natural objects. It is assumed that in addition to using fire to keep warm and to ward off wild animals, he may have cooked at least some of his food. In this connection there have been found in association with Neanderthal round balls of stone, and it has been thought that these may have been heated and placed in water as a cooking aid.

It is also thought that he lived in rough shelters or tents made of skins, and that he used skins as clothing to protect himself against the elements.

There is little evidence to show that Neanderthal man possessed any form of art or that he had any spiritual beliefs, but in some European caves of Mousterian times skulls of cave bears have been found associated with Neanderthal remains, arranged in a way that suggests a ritual of some kind. For instance, in the

Petershohle in the South German Alps a number of cave-bear skulls were set in niches in the cave walls. Some experts are of the opinion that there may have been a cave-bear cult equivalent to the bear cults existing among primitive tribes in the far north of Europe today.

It is also certain that Neanderthals had a great respect for their own dead, as skeletons have been found laid in graves shaped to fit the body, sometimes with a ring of animal bones surrounding them. In one case the skeleton still showed evidence that it had been covered with flowers. A curious feature of Neanderthal burials is that most of them show the corpses were oriented with the heads pointing to the west. This has been taken by some anthropologists to show that they were buried with a degree of ritual that can only be described as religious. It may indicate that Neanderthal had some idea of a heaven—comparable to the Egyptian concept of the land of the dead at the setting sun—in the west.

We can see that there are many peculiarities regarding Neanderthal man: the radically different types, which seem to vary between the "ape" and true humans; the fact that he used clothing and tools, possibly built shelters, and apparently had some form of religious belief and ritual. Is the picture presented to us by the anthropologists of a primitive "half man" a true one, we may ask? Is there another explanation for Neanderthal man?

Originally it has been suggested that Neanderthal man was stooped and walked on bent legs rather like a modern anthropoid. This view was modified by the hypothesis that the distortion of the limbs could have been caused by illnesses such as arthritis due to the cold and damp conditions under which he lived. Now it is thought that Neanderthal man in health was as upright as modern man.

Is it possible that *all* the deformities were caused by disease of some kind? Is what we are looking at a *distortion* of a true human being brought on by disease? We remember the large brain capacity of Neanderthal, larger than a present-day human's, in a creature whose bulk was the same as ours, or even less, for the average Neanderthal skeleton averages 5 feet, 5 inches. Even if this average height is deceptive because of bone malformation that gives an appearance of being shorter, the brain/body ratio is far greater than that of any animal except man.

The skeletons of what we term Neanderthal are surely those of true human beings severely affected by disease, and these distortions have mistakenly been taken as showing evidence of a separate, primitive offshoot of the human type. The enormous variation in skull type, from exceedingly primitive to almost those of Cro-Magnon, could have been malformations brought about by bone disease or genetic malformations by mutation. Remember, the remains we have discovered number barely sixty persons in all, and many of these are represented by only a few bones, or in some cases by just a part of a skull. Before we "invent" a new race or new races on the strength of such small evidence, we ought to question the existing hypotheses and suggest a new one.

We mention malformation by mutation, and this leads us to another idea, which has a connection with some peculiarities about Paleolithic man, and also with our next chapter.

Generally, today, we associate mutation with two specific causes—malformation of the genetic structure by chemicals (as in the case of Thalidomide), or by natural, or artificially induced, radiation.

We have suggested that there may have been a nuclear conflict in the past, and several writers have brought this concept to our attention in recent years. Von Däniken mentioned the use of nuclear weapons in his book *Chariots of the Gods,* and the theme was explored also in Tomas's book *We Are Not the First,* and by Kolosimo in *Not of This World.* Even as far back as 1909 Professor Soddy mentioned the possibility that nuclear science may have been known to a race so remote in time that it has left no record in history *(The Discovery of Radium,* Murray, 1909).

One reason for radiation damage in the past is the use of nuclear and possibly other weapons which brought about a change in the earth's orbit around the sun. If the earth had orbited in a more circular path, or slightly nearer to the sun, this would have meant a warmer planet, with the possibility of a higher rate of evaporation creating a high-altitude "blanket" of water vapor. This vapor layer would have had the effect of filtering out excessive radiation, and it is possible that the level of natural "background radiation" is higher now than it used to be in the past.

It has been suggested that one of the causes of aging is this continual bombardment of cosmic radiation experienced by all organisms. It is known that all radiation has a harmful effect on living organisms, and it is thought that this irradiation, although not deadly to life, is in some way responsible for the slowing down of cell replication, and natural wear and tear also plays a part. But radiation may play a major part in the aging process by slowing cellular replication in the first place and thereby allowing other factors to cause more damage than might otherwise have taken place.

A nuclear conflict in the past would have created a sudden upsurge of radiation levels which would eventually have exhausted itself, thereafter leaving a higher level of natural background radiation. When could this event have taken place?

We cannot say for sure, but we can make a guess to within several thousand years, taking into account certain factors.

It has been discovered that the cores taken from the bed of the Ross Sea in Antarctica indicate that glaciation commenced there some six thousand years ago. Vegetation of subtropical type examined in these cores died about this time, due to the climatic changes that had taken place.

Similarly, fossilized orange and magnolia trees have been discovered in the Arctic regions and appear to have lived in that area some thousands of years ago.

The commencement of civilization on this planet has been dated around 4000 B.C. in many parts of the world. This includes the Egyptian, the Sumerian, the Indus Valley civilizations, and possibly also some of the pre-Inca cultures of South America. Some Mayan remains in Central America have also been assigned dates almost as old as those of Sumer. All these civilizations have a tradition of great catastrophe, observed in Western cultures as the Flood.

It is likely, therefore, that the catastrophe, which some have equated with the termination of the last Ice Age, occurred between six and eight thousand years ago (6000–4000 B.C.).

According to this hypothesis, then, both Cro-Magnon and Neanderthal remains are not nearly as old as had been thought, but are more recent by many thousands of years. This is not to say that man has not existed here for many thousands of years, and the figure of thirty-five to forty thousand years for the existence of *Homo sapiens* may be almost accurate. On the other hand, man may have been here for tens or even hundreds of thousands of years longer than this, *in a highly civilized state,* the traces of which scarcely exist today.

Some mythologies (Egyptian and Sumerian) record that the Age of the Gods lasted for hundreds of thousands of years, which points to a tradition existing in antiquity of an enormously extended period of civilization in the remote past.

Neanderthal and Cro-Magnon remains have been found in association with the bones of the giant cave bear, the mammoth, the wooly rhinoceros, and the mastodons and other ancient elephants of Asia and America. As it is thought that these animals died out tens of thousands of years ago, it is therefore assumed that the remains of "primitive" man must also be of similar antiquity.

But is this really so? If all these species of animals, which are thought to be so ancient, were destroyed in a vast catastrophe, and this catastrophe occurred some six or eight thousand years ago, then our dates have been placed much too far back in time. A congress of scientists in the Soviet Union dined off mammoth steaks, for some of these creatures have been found frozen in the ice in a perfect state of preservation. Surely they could not have been there for forty thousand years? Also, representations of elephants have been found carved on stelae in the Mayan region of Central America, and it is thought in some circles that the American elephant may have vanished only during the past ten thousand years, with an occasional specimen surviving even into historic times (during the last two thousand years).

It could be true that all the skeletons we have found of primitive man, both Cro-Magnon and Neanderthal, living harsh lives in caves and by hunting, were the degenerate survivors of this great disaster, and that mostly they date from between six and eight thousand years ago.

We suggest, then, that what we have taken for a primitive form of man (Neanderthal) living both prior to and within the period of *Homo sapiens* was in fact the distorted mutation of true man himself. We suggest they were descendants of normal human beings living in highly irradiated areas following the holocaust, the victims of severe mutations caused by radiation. In time, of course, the most severely affected would die of radiation sickness, and there was probably also a high degree of sterility. Thereafter, as the radiation was dissipated, humanity slowly reverted to the norm, with the aberrant specimens dying out. *This would explain the disappearance of Neanderthal man.*

There is evidence of a use of nuclear weapons in the Americas as well as in Asia, and should we not therefore find traces of Neanderthal-type man in the Americas, if the causes of their being were radiation sickness? We have not found such traces of any apelike humans in the Americas. It may be that they are still to be discovered, or perhaps the survivors left the more badly contaminated regions before they could become so affected. There is a degree of evidence to show that traces of *Homo sapiens* in the Americas are as ancient as those in the Old World, so it is something of a mystery as to why there are no Neanderthal-type remains there.

We say that as radiation died down mankind reverted to the norm. But it was a norm that exists today, and *not* the norm that existed prior to the disaster. The norm *then* was an extended life span and a vastly superior intelligence.

Ancient man (whereby we include both Cro-Magnon and Neanderthal) generally had a larger brain than that possessed by man today, as we have seen. Of course, brain size in itself is no direct guide to intelligence: the degree of convolution, or folding, of the cortex is more important. However, from the shape and position of the skull features, it is thought that Cro-Magnon's brain was as highly developed as modern man's, as well as larger. This could point to a degree of natural intelligence higher than ours.

Some scientists have thought that since Cro-Magnon the human brain has been decreasing in size. Dr. Ernst Mayer has said that the trend may now be in a downward direction; that development of the brain stopped one hundred thousand years ago. This seems to point to a degeneration of the human species, and not, as is implied by evolutionary theory, to an advancement.

We are faced with this curious problem: If the human brain is in a developing state, why are there areas which have no known function? If we were developing, sections that are nonfunctional *would not exist,* and the brain would instead gradually enlarge over the generations as human intelligence increased. We have seen that the reverse is the case. Therefore, does it not seem more likely that the brain has to a certain extent atrophied, or lost much of its function, over the centuries?

We have observed another peculiarity regarding ancient man, and this is in the matter of trepanning, whereby a section of skull is removed either to examine the brain, to relieve pressure, or to remove a tumor or bone splinters caused by skull injury. Such operations today are extremely delicate and dangerous, and require the highest surgical skills. As recently as 1786 this operation, when carried out at the Hôtel Dieu in Paris, was fatal.

Yet Paleolithic man carried out a great number of such operations, judging from the skulls found which had been so treated. From the subsequent healing which had taken place it must be assumed that the individuals not only survived the operation but lived for many years afterward. When one considers that the number of human beings from the remote past whose remains we have discovered is remarkably small (all members of *Homo sapiens* found so far would not populate a small village), the number of trepannings discovered in proportion is extremely large.

It has also been discovered that a great many successful trepannings were carried out by pre-Inca surgeons in South

America at least twenty-five hundred years ago. Why did ancient man perform so many of these operations? We are given the usual explanations—it was done either to let out devils, or as part of some ritualistic magical ceremony, or even as a cannibalistic practice of eating the brains of an enemy. The last seems exceedingly doubtful in view of the number who survived the operation.

Of course, it is also admitted that some of these operations may have been carried out for medical reasons, since it is assumed that brain damage may have been frequent due to Stone Age man's habit of knocking one another over the head with clubs rather severely! It presents a mind-boggling picture that people who would indulge in such barbarous practices should also at the same time have surgeons capable of carrying out such· delicate and skilled operations. They must have had a more serious reason for carrying out such operations.

It is being increasingly realized that Stone Age man was nowhere near as primitive as he has been painted, and we will go so far as to say that there was in fact no such thing as Stone Age man. Just because people may have been thrust into a primitive environment by holocaustic events and have had to improvise with the most basic of tools, it does not necessarily follow that these people must have been primitive. The megalithic structures which abound throughout Europe, and which stem from a time far removed from those stages of civilization we term the Bronze and Iron Ages, represent a high order of intelligence and mathematical ability. In fact, since we are baffled as to how some of these structures were erected, they could point to a mastery of certain techniques of which we are ignorant today.

In a so-called burial barrow in southern England a set of metal implements were discovered recently which could not be readily identified. They were eventually found to be a set of surgical instruments closely resembling those in use today. Do these instruments stem from the period in which the barrow was built, or were they left there by later pople who used the structure? In some of these barrows there have been found the remains of fires, cooking utensils, and other household implements, so it looks as if their purpose originally was other than as burial grounds. They may have been shelters for the *living,* and if some of those living were injured it would explain the presence of the medical implements.

It also means that our divisions of Stone, Bronze, and Iron Ages are very loose descriptions and should not be taken too literally. After all, today we simultaneously have space rockets and Stone Age cultures, such as the Aborigines of Australia. In a thousand years' time, if the evidence of our civilization has vanished, and all that the new archaeologists discover are the stone tools of the Aborigines, they may well come to the conclusion that the twentieth century was a Stone Age period. Surgical instruments in the period we call the Stone Age may seem incredible, but surely far less incredible than the vision we have been given of a trepanning operation carried out on a human being with bits of flint hammering away into a human skull, without benefit of any anesthetic—and the patient often surviving afterward!

Taking the viewpoint of Stone Age man: if it were true that the brain capacity of man had started to decrease, and therefore also his intelligence, the people who carried out all these trepannings—these people who were the survivors of a highly advanced civilization—would want to know why. They may have wondered why their memories were failing; why they were unable to carry out mathematical calculations with their former precision; or why they were subject to severe head pains and tumors. Were they, in fact, carrying out these numerous operations to try to discover what was going wrong, and perhaps also to remove cancerous tumors?

Radiation sickness has been frequently described: the loss of hair, lassitude, vomiting, weakness, and eventual death. It is also known that the first of the cellular structures to be affected are the brain cells. If there were a high level of radioactivity due to nuclear fallout, it is possible that there could have been a high incidence of brain damage, coupled with a loss of faculties.

Further, the increased radiation could have been the key factor in lessening the life span of the human race. Thus we observe in Genesis that after the Flood there was a systematic decrease in longevity.

The possibility of a high radiation level at a particular time in the past brings us to another point. It was suggested by L. A. G. Strong in his book *Flight to the Stars* that perhaps the various monsters encountered by the ancient Greek heroes such as Jason —remember Cyclops, with one eye in the middle of his forehead— were actually mutations caused by radiation from a starship's engines. Strong hinted that there was a possibility the gods and monsters of Greek myth actually were spacefarers visiting this planet (anticipating Von Däniken!) and that mutations may have occurred as a result of a too-close proximity to the ship's atomic-powered engines.

Perhaps there is a more terrestrial answer: that such mutations were a result of a nuclear holocaust on earth. We are reminded that our mythology and traditions are peopled with many odd monsters: ogres and hairy submen, people with horns on their heads, satyrs with the feet of goats, werewolves, vampires. Although these legends are difficult to trace to their source, many seem to stem from a particular point in history, in the same way that all the world's major religions seem to have come into being at the same time. Could it be that all these varied monsters, both human and animal, were in fact the victims of massive exposure to radiation?

Of course, such creatures no longer exist, and so the traditions are dismissed as fairy tales. But, like the birth of the gods, *something* must have triggered off such ideas. There seems no reason, *at the source*, why such tales should have been invented. Perhaps such creatures did in fact exist but do so no longer because they eventually died out, as do all mutations unfitted for survival. As the race reverted slowly to normal, it left behind a memory of distorted creatures lingering in the uneasy subconscious of mankind.

5

A Nuclear War—
5000 B.C.

A NUCLEAR WAR some six to eight thousand years in the past? A fantastic idea—an impossibility? This idea, which would explain many mysteries about our past, and which at one time would have been dismissed out of hand, is being considered seriously today. We have heard it said by laymen—people whose minds are uncluttered by dogmatic scientific ideas—that an atomic war in the past would explain many things not at present understood. Perhaps the layman has hit on truths that are obscured by the scientific mind's overrationality.

Before the dawn of our own atomic age many puzzling features of the earth, and some aspects of mythology, would not have been linked with such a postulate simply because there was no conception of atomic weapons and their capabilities. For instance, until we started to experiment with radioactive substances, no living person on this planet could have described radiation sickness, for the simple reason that such a disease *did not exist*, and therefore could neither be predicted nor described.

Yet radiation sickness, in clinical detail, is described in an ancient Sanskrit text, the *Mahabharata*. Such an affliction could not have been described unless it had been experienced; otherwise how would the ancient chroniclers have been able to picture it so precisely?

The *Mahabharata* describes clearly the explosion of great fireballs, the gales and storms which they created, and the after-effects. These aftereffects took the form of hair loss, vomiting, weakness, and eventual death: classic symptoms of radiation poisoning. Even more significantly, it was said that persons in the vicinity of these weapons could save themselves by removing all metal from their persons and immersing themselves in the water of rivers. There could be only one reason for this—to wash away contaminated particles—the same procedure followed today.

There seems to be a large body of evidence showing that nuclear science was known in the past. There is the *Mahabharata*, frequently cited by modern writers. Von Däniken mentions the description of nuclear weapons from this ancient text, and his explanation is that they were used by advanced aliens from space against the primitive people of this world. But it does seem unlikely that an advanced race would take such measures against primitive people armed only with spears and bows and arrows.

The *Siddhanta-Ciromani*, a Brahmin book, subdivides time until it reaches a final unit, truti, which is 0.33750 of a second. Sanskrit scholars are puzzled as to why such a small unit of time was used in antiquity, or how that unit could have been measured without instrumentation. Modern primitive peoples are notoriously vague about time—even hours have little meaning for them. There is no reason to suppose that ancient primitive people should have viewed time any differently, which makes the subdivision even more puzzling.

Tomas, in his book *We Are Not the First* (Souvenir Press, 1971), says, in the chapter "From Temples and Forums to Atomic Reactors":

"According to Pundit Kaniah Yogi of Ambattur, Madras, whom I met in India in 1966, the original time measurement of the Brahmins was sexagesimal, and he quoted the Brihath Sakatha and other Sanskrit sources. In ancient times the day was divided into 60 kala, each equal to 24 minutes, subdivided into 60 vikala, each equivalent to 24 seconds. Then followed a further sixty-fold subdivision of time into para, tatpara, vitatpara, ima and finally kashta—or 1/300 millionth of a second. The Hindus have never been in a hurry, and one wonders what use the Brahmins made of these fractions of a microsecond. While in India, the author was told that the learned Brahmins were obliged to preserve this tradition from hoary antiquity, but they themselves did not understand it."

The time unit of kashta—1/300 millionth of a second—is absolutely meaningless without instrumentation and, more significantly, is close to the life spans of certain hyperons and mesons —*atomic particles.*

The Varahamira Table, dated approximately A.D. 550, gives a mathematical figure that compares closely with the size of the hydrogen atom. Were these figures also handed down from a much more distant time?

The yoga Vasishta says: "There are vast worlds within the hollows of each atom, multifarious as the specks in a sunbeam." This seems to hint at the knowledge that not only is matter made of numberless atoms, but that the atoms themselves are, as we know, mostly *empty space.*

These writings, which stem from a remote period, suggest that knowledge of atomic physics existed in the past. The fact that the Brahmins were *obliged* to remember these mathematical symbols, even though they did not understand them, represents an effort to transmit knowledge from a vanished technological era. One can imagine scholars, upon observing their civilization collapsing, writing down their knowledge and entrusting to a certain group the responsibility of passing the information down the centuries, until the time should come when it would be understood once more.

Much has been lost; what has survived has survived in only a fragmentary form. But hidden in monasteries or in obscure texts, which have never been brought to the attention of scholars, there may still exist much more information about nuclear physics. As it happens, we have discovered the nuclear age for ourselves by a separate route, although it might have been better if such knowledge had never been rediscovered.

Something of this ancient knowledge seems to have filtered down through the ages in a more generalized form. Democritus two and a half thousand years ago proclaimed: "In reality there is nothing but atoms and empty space." The Greek thought the atom was the smallest unit of matter and could not be divided, whereas Moschus, the Phoenician, who had acquainted the Greek with this information, asserted that the atom could be divided.

Lucretius, a Roman scholar of the 1st century B.C., wrote that atoms "rushed everlastingly through space, and underwent myriad changes under the disturbing impact of collisons. They were too small to be seen."

After the collapse of the Roman Empire it was not until the nineteenth and twentieth centuries that serious work on atomic physics was undertaken.

Of course, the ancient scholars, whose fragmentary knowledge must have stemmed from a remote and forgotten technical age, could not demonstrate their theses in practice. The technology did not exist whereby they could do so.

A remote age during which nuclear science was practiced implies the use of atomic energy for many purposes. Some ideas, such as transmutation, which the alchemists kept alive in their endless search to turn lead into gold, most likely stemmed from ancient knowledge that manipulation of atomic structures could convert one element into another.

However, what we are concerned with in this chapter is not the use but the misuse of nuclear energy. It seems clear that a frightful holocaust occurred in a past age, and writings from different sources confirm this. Whether the conflict was purely terrestrial or involved another advanced race from elsewhere in the universe is a matter of speculation.

From India we have the evidence of the *Mahabharata* and the *Drona Parva*, which speak of great fireballs. Kapilla's glance, which could burn fifty thousand men to ashes in seconds, sounds like either nuclear energy used as a beam, or some kind of laser weapon. The *Drona Parva* speaks of flying spears that could ruin whole "cities full of forts."

Flying spears that could ruin cities—could these be missiles, possibly with nuclear warheads?

India is not alone in legends such as these. They exist also in China. Raymond W. Drake says that the accounts in the *Feng-shen-yen-i* describe events with a close similarity to the Sanskrit *Mahabharata*. Rival elements fought for control of China, helped by celestial visitants who used weapons that remind us of our own advanced technology. The war was fought with blinding rays and dragons of fire, spheres of flame, shining darts, and lightning. According to the descriptions they seem to have possessed something akin to radar, whereby they could see and hear objects many hundreds of miles away. There were flying dragons of silver, and chariots of fire and wind. These all seem to refer to laser beams, nuclear weapons, missiles, and flying machines using rocket or jet engines.

In Siberian shamanistic legends there is a warrior with dazzling arrows who blows up anyone who laughs at him and rides away through the sky on a "shell of gold."

The ethnographer Baker was told by a Canadian Indian, a wise man of a secret totemic society, that once there were great forests and meadows "down there" and also great shining cities and men who flew in the skies to meet the thunder bird. Then the demons came and the cities were all destroyed and only ruins exist now.

These legends stem from the regions of permafrost in the far north of the Canadian tundra, and they must refer to an ancient epoch when the climate in the polar regions was very different from what it is today. "Down there," we must assume, was to the south where the shining cities lay.

This Canadian Indian tradition should remind us of the legend existing among the Maya and Aztecs about the cities where the lights never went out by day and night.

The Maya say: "These lands [referring to the southwestern part of what is now the United States] are the Kingdom of Death. Only the souls that will never be reincarnated go there . . . but it was inhabited a long time ago by ancient human races."

There is a great similarity between the legends of a holocaust from many different parts of the world—consider Zeus of the Greeks with his lightning bolts and the Nordic Thor with his hammer and lightning. If there had been a world conflict in the remote past, wider even in scope than World War II and much more devastating, then surely this is what we should expect. Global conflict would leave echoes in global memories.

In the same way, the legends of the Golden Age are equally widely dispersed. Those who seek a marvelous "Atlantis" in any one place are, it seems, doomed to disappointment. If the whole world was highly civilized at a certain period in the remote past, then the legends of its former existence would be worldwide in extent. The fact that different accounts from all points of the compass have led people to believe that Atlantis existed in the Mediterranean, the Atlantic, Spain, Greenland or Iceland, America or Tibet, or even that an Atlantislike kingdom of Lemuria existed in the Pacific, merely points to a civilization that was worldwide. This civilization was not limited to one mysterious island or continent now sunk beneath the sea.

A global civilization at a *remote* period would explain the similarities between widely separated cultures of antiquity—*and also the differences*. There are broad points of similarity between the civilizations of Egypt and Sumer, and between the Indus Valley and those of Central and South America, particularly in their legends. There are also great differences, although these may be largely a matter of *detail*, such as the Old World possessing the wheel and the New World not making use of this device. It should be remembered that at the period of megalithic building *all over the world*, the wheel was not in use anywhere.

Those who would completely debunk the idea of there being any connection *at any point in time* between all the civilizations of the past, both in the Old World and the New, do not seem to be taking all the factors into consideration. However, we seek in vain for one isolated area as the home of our "mother culture" from which all civilizations descended. The whole earth was the "mother culture."

We may note a further peculiarity. Many of the legends that

point to a great and disastrous conflict in the remote past point also to the regions where these conflicts took place. *And most of these regions are now deserts.*

The Chinese have such legends, and in that great land lies the largely unexplored Gobi Desert, which hides mysteries beneath its sands. India has its deserts, and where the Indus Valley civilization flourished is now desert. Much of the southwestern United States is desert, and the Maya called that the Kindom of the Dead. In Egypt Horus cursed the Lands of Set for thousands of years, and vast tracts of North Africa, including the great Sahara, are deserts. Much of the Middle East is desert, with the great cities of the Babylonians and Sumerians lying now as ruins under the shifting sands. Australia is largely desert, and the Aborigines would seem to be the descendants of once highly civilized people. Perhaps here also lie buried the traces of cities from long ago.

Legends, mythologies, accounts of ancient weapons that sound like nuclear missiles, others that seem to refer to the mathematics of nuclear physics. The evidence seems to be piling up in favor of our hypothesis. Do we, however, need more than old texts and legends—more concrete evidence to support our views? This material evidence also exists.

When nuclear weapons were test-fired in the New Mexico desert, and also in the Gobi Desert near Lop Nor, the sand was vitrified (fused into glass) by the heat of the explosions. Intense heat is needed for this—volcanic action will not produce this effect, nor ordinary explosives, nor fires. The heat of millions of degrees, that of thermonuclear reactions, is necessary.

The Gobi Desert has areas of vitreous sand that have been there for thousands of years. Similar ancient glassy sands exist in parts of the California desert, and the mysterious glassy tektites found in the sands of the Middle East and North Africa are thought to have their origin in radioactive processes. Was it thermonuclear heat that produced these ancient areas of glassy sand? Rocks in Peru and Bolivia show evidence of vitrification, and desert conditions exist there also.

Should we not ask ourselves, Why do these great deserts exist? What brought them into being in the first place? Why should there be deserts surrounded on all sides by huge areas of vegetation?

Could it not be that these deserts were created in the first instance by the exploding of nuclear missiles of great power, destroying not only those people who may have lived there but all life as well, with radioactive contamination sterilizing these areas for centuries to come? Little grows in these deserts, and all that grows and lives there has developed special adaptations to the harsh conditions and survives only with difficulty. Perhaps there is truth in the legend of Horus: the curse against the Lands of Set was the curse of radioactive poisoning.

It could be said that our hypothesis is faulty if there were no trace that humanity had lived in these now-barren regions, that they had from time immemorial been lands hostile to large-scale human settlement. But this is not so.

There are ruins in the Gobi Desert, almost formless and of great antiquity. These ruins bear the marks of blistering by great heat, of the same kind that were noted at Hiroshima.

Some geologists maintain that many of the present desert areas of the southwestern United States and their curious rock formations were not brought about by natural causes.

During excavations near Chillicothe, Illinois, bronze coins were dug up from a depth of more than 127 feet. The pressure and the vast period of time which has elapsed since these coins were buried has erased all traces of any markings they may once have held.

In 1851 in Illinois two copper rings were found at a depth of more than 109 feet. In June of that year a vessel was found encased in pudding stone at Dorchester, Massachusetts. It was a bell-shaped jar made of an unknown metal with a floral design in silver.

A doctor in California found, inside a piece of gold-bearing quartz, a small gold object resembling a bucket handle. A similar handle was found in a cave in Kingoodie, northern England, in a block of stone, and it has been estimated as being at least 8,500 years old. One can assume that the California find is of similar antiquity.

In 1969, inside a rock from the "Abbey Gallery" in Treasure

City, Nevada, were found the traces of a screw two inches long. The screw had long since disintegrated, but the space it had occupied left its trace in the shape of the spiral thread markings. This was possibly tens of thousands of years old.

An American farmer named Tom Kenny, in the Plateau Valley, discovered, at a depth of more than ten feet, a section of pavement made of smooth symmetrical tiles. Analysis of the mortar proved it to be unlike the local valley materials.

All these things point to a possibility not previously considered: that there exist traces, sometimes buried far underground, of a civilization many thousands of years old in *North* America, and that this civilization appears to have been violently destroyed in a vast catastrophe involving levels of heat which can only be attributed to thermonuclear explosions.

Of civilizations in Central and South America we have evidence in abundance. Were those who founded the earliest elements of these civilizations survivors of the vast cataclysm to the north? The original habitat of the Toltec and pre-Toltec peoples, who built the vast pyramid complex at Teotehuacan, are not known; and the Aztec had a tradition of traveling down from the north—from the "Country of Painted Colors," possibly the Grand Canyon region with its bands of different-colored rocks.

The pre-Inca and Inca peoples of South America worked in gold, silver, copper, and bronze—no other metals. Yet in the sixteenth century Spanish conquistadores found a seven-inch *iron* nail inside a rock in a Peruvian mine. The rock was estimated to be tens of thousands of years old. Who had made this nail so long before the time of the Incas?

Ancient iron nails have turned up elsewhere. In California, a Mr. Hiram D. Witt in 1851 discovered an iron nail with a perfect head inside a piece of auriferous quartz. The discovery was made, according to the London *Times* of December 24, 1851, when he dropped the quartz, which broke and revealed the nail. Again, in Kingoodie Quarry, where the golden "bucket handle" was found, an inch of iron nail, including the head, was found embedded in rock. All these items stem from a period before metals were used at all, according to the experts.

China, the Americas—here exist the legends of a vast conflict, and here also the evidence. In India, too, where we have the clearest reports of a past high technology, there are strange things.

The explorer De Camp mentions the existence of charred ruins between the Ganges and the mountains of Rajmahal, which seemed to have been subjected to intense heat. Huge masses had been fused together and hollowed "like lumps of tin struck by a stream of molten steel." A British official, J. Campbell, observed similar ruins farther to the south. Other travelers have described ruined buildings made of unusual materials "like thick slabs of crystal," and these also had been subjected to extremes of heat, being holed and split by enormous powers. A skeleton was found in India with a radioactive level fifty times above normal. (A. Gorbovsky, *Riddles of Ancient History,* Moscow, 1968.)

Buried under the shifting sands of the deserts there must be many things formerly unsuspected. Perhaps it would be useful to check in these areas for traces of higher than normal radiation. Perhaps the ancient radiation has long since dissipated, but there still may be lingering traces.

The jungles of India and Central and South America have scarcely been investigated. The subcontinent, for instance, has areas that have never been thoroughly explored. There are remote villages no white man has ever seen, in spite of the fact that Europeans have lived and worked in India for several centuries, building railways and dams and modern cities. These mysterious regions may hold keys to the past.

Do the pyramids of Egypt hold a clue to the riddle of a nuclear holocaust in ancient times? The Ein Shams University of Cairo, under Dr. Luis Alvarez, placed a cosmic-ray detector in the base of the pyramid of Khephren for the purpose of ascertaining if there were any chambers in the pyramid which had not been discovered. Cosmic rays shower down uniformly, and if there were any unknown hollow areas within the pyramid, the cosmic rays would pass more easily through these spaces and leave heavier shadow traces on the detector.

In September 1968, tests showed the detector to be in perfect working order, and yet the hundreds of recordings made during 1967–69 that were analyzed by the IBM 1130 computer at the university showed *no common daily features*. Dr. Amr Gohed, in charge of the installation, stated: "This is scientifically impossible. There is a mystery which is beyond explanation . . . there is some force that defies the laws of science at work in the pyramid."

In *Colony: Earth* we suggested, being guided principally by legends—there is no other guide, as the ancient Egyptians themselves regarded the pyramids as a mystery—that the pyramids were not tombs but shelters constructed in the event of a vast catastrophe; not necessarily shelters for human beings, although some may have been sheltered there, but perhaps for the safekeeping of information and knowledge.

If the catastrophe took the form of a man-made disaster, including the use of nuclear weapons, then this hypothesis is given added weight by those cosmic-ray tests and their strange results. It may be that the pyramids, in addition to their other attributes as impregnable shelters, were also *radiationproof*. It also appears that whatever was used for this purpose *is still functioning*. Is there, as Tomas suggested, a generator situated somewhere beneath the pyramids? (See *We Are Not the First,* by Andrew Tomas, Chapter 22.)

Such a device may be situated deep underneath the pyramid, far below any excavations which have been attempted so far. What form it may take or how it may be detected is difficult to say, because such a device is beyond our technology, and we do not know on what principles it may operate, how large it may be, or the way it is powered.

Indeed, there are still many unsolved mysteries about the pyramids, and yet there are those who still insist that they were merely tombs of the pharaohs. This is not to say that we agree with the more fanciful stories that have been developed regarding these structures. There are some who say that the height of the Great

Pyramid of Cheops, multiplied a million times, shows the distance from the earth to the sun. In fact, various measurements and figures have been produced from time to time to show all sorts of things—the length of the year, the lunar month, the weight of the earth.

Some have suggested the pyramids were built by Noah after the Flood, or that they were built by spacemen, or on the instructions of spacemen, or that they were guidance markers for astronauts from other planets.

The idea that all the world's knowledge is concealed somewhere within the pyramids may not be altogether fanciful, but it is probably truer to say that the knowledge possessed by the ancient world *was* placed in them, but is no longer there. So much mystique has grown up around the Egyptian pyramids, particularly in the last hundred years, that the truth probably lies between the outrageously far-fetched and the prosaic.

We cannot agree with the more wild theories, but neither can we agree with the idea that these tremendous structures were built by a group of people without the wheel, with few tools, and no hoists or other weight-lifting tackle. Or that this was done in a country which, at the time the pyramids were supposed to have been built, was almost uninhabited (the population in total is reckoned to have been some 2 million), and possessed no cities and scarcely any trace of other than small settlements.

We are of the opinion that the pyramids, particularly the vast group on the Giza plateau, are pre-Egyptian. They were not built as religious structures or tombs but were protective devices, possibly for the storage of knowledge, much of which was used by the "culture bearers" when they emerged or returned to lead the savage descendants of the survivors of the catastrophe back to civilization.

Of all the pyramids in Egypt—and there are over sixty known—there has not in a single one been traced beyond doubt any burial. All the pharaohs, especially those most important and powerful rulers such as Rameses II and Thutmosis III, were buried in the famous Valley of the Kings. This is where the treasure of the boy king Tutankhamen was found, we remember. It is surprising how many people think that the mummified remains of the pharaohs were discovered within the pyramids, simply because it has been stated so categorically by the experts that the pyramids were tombs of the rulers of Egypt.

If the greatest and most illustrious rulers of Egypt were buried in rock tombs in a valley, why is it assumed that less important, obscure, and by no means always real kings (the ancient royalty lists of Manetho do not have a Cheops) were buried in these vast edifices supposedly raised by so much human toil and sweat? The Egyptologists appear to be largely influenced by the records handed down by that great traveler of antiquity, the restless Herodotus. But Herodotus was guided by what the scholars and priests of his day told him, when the pyramids were already thousands of years old and as much a mystery to the Egyptians of the period as they are to us today.

The purpose of the pyramids surely must be far different from that which has generally been supposed.

In the preceding pages we have discussed the possibility and offered the evidence for a nuclear war in the past. We have seen that not only nuclear bombs may have been known but also missiles and something akin to radar, together with aircraft. None of these things could have been produced without the resources of a highly advanced technology and therefore an industrialized society.

This being the case, the ancient lost Golden Age was probably only a Golden Age in retrospect—to the survivors, who probably remembered the benefits and preferred to forget the disadvantages.

In reality this lost technological era may have suffered from all the ills which beset ourselves, and, indeed, if its people were so much more advanced than ourselves, their problems may have been that much greater. The fabled Cities of Light must also have had their slums, traffic, noise, and stress; they also may have had labor and industrial problems; possibly difficulties with vast populations—overcrowding and the dangers of breakdown of the food supply. Even if these people were truly the godlike figures of mythology, with marvelous intellects and long life spans, there could still be psychological flaws in their makeup that could have created the same dangerous situation we find ourselves in today in our society.

Or if they were highly advanced colonists from another solar system, the radiation from the sun may have taken its toll. Possibly they suffered a degree of degeneration over the ages—we have mentioned the fact that radiation affects the brain structure more quickly than other parts of the organism. By the time a purely terrestrial conflict was imminent they may have already started to degenerate, even though remaining immeasurably superior to ourselves. This aspect would explain the Genesis statement that man became corrupt and that "evil was in his heart continually."

On the other hand, they may have perfected a society far more advanced and more humane than ours. Perhaps the fault lay in their relations with other races elsewhere in space, and the catastrophic conflict that wrecked their world and almost destroyed the human race was brought about by a hostile intelligence from another planet or an invasion from another solar system.

Perhaps it was a combination of all or some of these factors. It may be that we shall never know the exact circumstances of man's "fall" but will have to rely on our interpretation of the mythologies and the available evidence.

It does seem certain that something is sadly amiss in our planetary system, if planetary systems are formed in the manner described by our astronomers. We shall go into this point in detail in our next chapter.

Many interesting speculations can be made—and supported —on the possible techniques of the great conflict. If a past era with a high technology possessed nuclear weapons, missiles, radar, etc., if its people followed a parallel to our own pattern of development, they will also have possessed other methods of waging war. We have evidence from both mythological and material sources that this may have been the case.

An ancient Sanskrit text, the *Samara Sutradhara*, talks of the development of *biological* weapons in the remote past. Samhara was one which crippled, and Moha induced paralysis. The *Feng-*

shen-yen-i mentions biological warfare used in ancient China, and again we have a close similarity to the descriptions from ancient India.

This could lead us to an interesting thought: Is it not possible that certain diseases may once have been artificial in origin? There are many that seem to be exclusive to the human race, which do not affect animals. Venereal disease seems to hit only humans, and although indiscriminate sex is blamed for its spread, many species of animals are far more indiscriminate in their sexual relations. Cholera and typhus, although carried by animals, do not seem to affect the carriers. Could these and other human afflictions have been artificial in origin, the result of ancient biological warfare, raging unchecked, when barbarism descended on the world and medical services collapsed, and man had to await the coming of our twentieth century to once again find the cures?

It is an interesting thought, because Firsoff has pointed out the fact that the virus, now regarded as the intermediate stage between life and nonlife, may not be ancient after all. Many biologists point to the virus and say that here we have the prime example of the first stage of the division between the organic and the inorganic world. The virus in its inert state behaves like a crystalline formation (the inorganic) and when active it replicates and behaves in a purposeful manner (the organic). Firsoff has mentioned the high degree of "host specificity" of the virus, which could point to its being a recent development.

In other words, some viruses have a high degree of host specificity to the human being, and as man is reckoned to be a recent arrival on the planetary stage, the virus also can be regarded as recent. We can go a step further and suggest that perhaps certain viruses were originally "tailored" to attack human beings by being artificially created in the laboratory. Viruses have recently been synthesized in our modern research laboratories by recombining existing virus material, thus producing new strains not found in nature. Therefore it is not impossible that a highly advanced society in the past may have done a similar thing. This would connect closely with the reports of biological weapons in ancient documents.

We will now consider another aspect of ancient weaponry which also exists in our present-day society: projectile weapons in the shape of pistols, rifles, etc., and high explosives.

A human skull was discovered in a cavern in Northern Rhodesia and is now on display in the Museum of Natural History in London. There is a perfectly round hole in the left side of the skull and the right side is shattered. There are no radial marks—the hole is perfectly round—and this pattern of a perfectly round hole with the opposite side shattered is typical of the effect of a high-velocity rifle bullet. A spear or arrowhead would not cause such a hole, and neither would create the shattered effect on the opposite side. Only a hot, high-velocity projectile can do this. This skull has been estimated as being *forty thousand years* old! Yet we are told that man, at that date, had only just appeared in the world and had not yet even invented the bow and arrow.

This skull is not the only thing that shows such a peculiar injury. The Paleontological Museum of the U.S.S.R. possesses a skull of an auroch which is many thousands of years old. An auroch is an extinct type of cattle which was supposed to have vanished from the earthly stage tens of thousands of years ago; yet this ancient animal skull has a perfectly round hole in the front of the forehead. Examination has shown that the creature, although injured, was not fatally so, and the wound had healed. Like the Rhodesian human skull, the hole was perfectly round with no splintering effect, in the manner made by a rifle bullet.

Rifles in the remote past—could it be possible? We have the concrete evidence, and we are reminded of a frequent theme in mythology: rods that can spit fire. There are myths from Druidic times that talk of rods which spit fire and could kill. Moses had a rod that brought water from a rock when he struck it. It *could* have been a divining rod—water divining is an ancient art—but it might have been a projectile device.

We know that Moses was brought up from babyhood as a member of the Egyptian ruling class, and he may have had access to knowledge restricted to the rulers and the priestly class. It is somewhat of a mystery how the ancient Egyptians managed to drive tunnels through solid rock, as they did in the Valley of the

Kings and in constructing the Abu Simbel "House of Eternity," which Rameses II had built into a cliff face. It has been suggested that in the earlier periods of Egyptian history there was a knowledge of explosives that was later lost. This is not as strange as it may seem—the Chinese were making explosive powders several thousand years ago, although they did not put it to any practical use. The possession of firearms in remoter periods is not unbelievable, therefore, and to a civilization that possessed nuclear weapons, the use of rifles and other similar weapons should not seem at all remarkable.

An even more remarkable suggestion has been made by some Soviet scientists. Some discoveries of the skeletons of dinosaurs seem to point to the fact, judging from the nature of the fractures and the position of the skeleton bones, that they were shattered by the use of high explosives. This, of course, would place the use of explosives (by someone) many millions of years in the past, and here we are reminded of an earlier suggestion in this book, which is that the dinosaurs may have been deliberately destroyed by man either before or during colonization of the planet.

Clearly, the knowledge of and use of explosives, and weapons based on explosives, is immeasurably more ancient than we have always supposed.

Finally, we shall turn to another mystery that may well have a connection with a nuclear holocaust in the remote prehistoric past, and this is the matter of certain cave paintings and rock drawings scattered around the world.

This concerns the so-called spaceman figures which have received considerable publicity in recent years. The vogue was first given full prominence by von Däniken in his book *Chariots of the Gods*, but it has also been mentioned by many other writers.

The best-known drawing is the giant outline carving on the Tassili Plateau in the Sahara, discovered by Henri Labote and christened by him the Great Martian God. It could hardly have been called anything else—the resemblance to a figure in a space suit is quite remarkable. There are other, very similar, drawings in the Tassili region: one showing a group of four figures walking, who appear to be wearing something like space suits, complete with bulbous helmets.

Drawings like these have been found in many parts of the world. There is a rock drawing south of Fergana in Uzbekistan (U.S.S.R.) that shows a figure with the head enclosed in a ring with rays coming from it, which could represent a transparent "fishbowl" helmet with antennae. There are almost identical figures from Val Camonica in Italy. "Spacemen" figures have been found drawn on rock faces in Australia.

Related to these are the Dogu statuettes from the Jomon period of Japan, which have aroused considerable interest. These statuettes represent people wearing protective clothing of some sort, together with helmets having curious eyepieces. Isao Washio, the Japanese expert on these figures, says: "The gloves are fixed to the forearms with a rounded attachment while the eyepieces can be opened or closed. There are levers at their sides perhaps meant for manipulating them, while the 'crown' on the helmet is probably an antenna . . . the designs on the suits are not ornamental but correspond to devices suitable for regulation of pressure automatically."

All these drawings and figures are today associated with spacemen and with the possibility that this planet has been visited by astronauts from other worlds in times past. They may, in fact, be representations of our early "sky gods." The fact that these figures are often associated with flying discs, with spheres in which they are sometimes enclosed, and with other flying devices would appear to add weight to this theory. At an ancient astronomical observatory at Meroe there is further evidence of such flight: a perfectly clear drawing of something that appears to be a tapered rocket or missile, complete with antenna.

While it is agreed that there is a good case for these figures representing spacemen, there could also be another explanation. Are they, perhaps, people wearing protective clothing to guard them against radioactive contamination?

We observe that many of these drawings of the "spacemen" images have been found in present *desert* areas. We have already suggested that perhaps the deserts were created in the first instance by the use of nuclear weapons, therefore these regions would still be dangerously radioactive thousands of years later. The pictures could have been drawn by surface-dwelling survivors who had regressed to savagery; the suited visitors on survey missions had arrived by some sort of flying machine, either from areas not so severely affected by contamination, or from their deep underground shelters. We are aware that there exist many legends of man's possession of flying machines in the past Golden Age; therefore the association of these suited figures with aircraft does not necessarily mean that they are extraterrestrials.

Nor do legends of flying machines exist only from the remote past. In 1972, near the Step Pyramid in Saqqara, there was discovered under the sand a wooden model of what appears to be an aircraft, very similar in its design to modern jet aircraft. The object has been dated as in excess of 5,500 years old. One daily newspaper dubbed it as "Tutenconcorde," and although most experts thought it was merely a rough model of a bird, this seems unlikely. The ancient Egyptians were perfectly able to depict the shape of birds—the most popular one being the emblem of Horus, a hawk—and this object looked exactly like a rough wooden representation of an aircraft, the sort of thing someone not very good with his hands may make for a small child, or the kind of thing a child himself would make.

Of course, some of these drawings may represent visitors from other worlds, but we cannot overlook the possibility that these protective suits were worn by terrestrials as a shield against radioactive contamination. If people from other planets had to be so thoroughly protected against the terrestrial environment that they wore fully insulated spacesuits, it would most likely mean that they would be unable to breathe our atmosphere or tolerate our temperatures, which does not fit in with our picture, from mythology, of the sky gods, either in appearance or behavior.

If it is assumed that the ancient gods were spacemen, they must include Wirakocha, Quetzalcoatl and Kukulkan of the Americas, and perhaps also the gods of the Egyptians, Greeks and

Hindus. These are never described as wearing protective suits. Therefore they were either terrestrials or aliens from other worlds whose physiology was so similar to that of terrestrials that they did not need space suits. Of course, short-stay visitors may have worn space suits as a protection against unfamiliar solar radiation or against unknown and possibly highly dangerous terrestrial microorganisms.

However, on balance, considering the location of the principal drawings and the evidence offered for a past nuclear holocaust, protection against radiation hazards may seem the more reasonable explanation.

6

Clash of Worlds

IT IS NOW generally assumed by astronomers that solar systems are the rule rather than the exception where solar-type stars exist in the universe, and that there are possibly many millions of planetary systems in our galaxy alone.

The reason for this opinion, as against the earlier one which held that solar systems were extremely rare, is that newer concepts have stemmed from a better understanding of the nature of the universe. At one time it was thought that a wandering star pulled a filament from the sun as it passed close by, and that this filament later condensed into the planets and moons of our solar system. This line of reasoning meant that, as such near-collisions were extremely rare, planetary systems would also be correspondingly rare.

However, the general design of our planetary system does not conform with such a theory, and it was replaced by the theory of the condensation of a cloud of gas and dust—the "cold accretion theory." This held that a cloud of dust and gas would condense into globules and rotate around a center of gravity by centrifugal force; the central part, or core, forming the sun, and the condensation of the outer parts forming the planets and moons. This proto-system, with its whirlpool, *almost* fits the present shape of the solar system and, although not altogether satisfactory, is the most reasonable of all the alternatives offered to date.

96

Basing their reasoning upon our solar system, astronomers have devised a model for a "life-bearing zone" around solar-type stars for the planets orbiting their primary; the zone depending on the size and radiation emission of the star. In our planetary system the zone in which life as we know it is possible, could theoretically extend from the orbit of Venus to slightly beyond Mars. Admittedly, Venus would be hotter than the earth, and regions as far as Mars and beyond somewhat colder, but not so much so as to exclude the possibility of life resembling that of earth—provided that these three planets had developed in a similar manner.

According to more recent theories of planetary origin and development, their composition—originating as it has from the same gas and dust cloud—should have been almost identical, and their subsequent evolution should have followed similar paths. Venus is almost a twin of the earth in size and gravity, and Mars, although smaller, has a gravitational field of sufficient strength to have retained a considerable atmosphere. However, the three planets have, in fact, been shown to have vast differences.

The Venus probes sent by both the United States and the Soviet Union have presented us with a picture of a planet of extreme surface heat—temperatures in the region of hundreds of degrees centigrade—so hot, in fact, that some experts have said that the surface probably has a faint glow. These surface temperatures are sufficient to melt softer metals, and probes soft-landed on the surface have ceased to function almost immediately. It is extremely rare that we are able to see any surface markings at all on Venus, due to its dense cloud cover, but recent radar echo-sounding techniques suggest that there are at least two long and extremely high mountain chains on the planet.

At the present time the most popular model for this planet is that of a hot, waterless desert torn by tremendous gales originating in a dense atmosphere of carbon dioxide. Any water vapor is concentrated in the upper layers of the atmosphere and is in the form of ice crystals. This extreme difference between the cold of Venus's upper atmosphere and the tremendous heat on the surface is but one of the planet's many mysteries. The surface atmosphere of carbon dioxide appears to make it certain that life, particularly vegetable life, does not exist on Venus.

Another model which has been suggested, although there is as yet no concrete evidence from the Venus probes to support it, is that the surface may have a high hydrocarbon content in the form of petroleum. A large part of the surface may be covered in petroleum oceans.

Measurements of Venus's movements have provided two surprises. One is that the planet's rotation is extremely slow—each "day" lasts 243 days. The other is that it rotates in a manner opposite to all the other planets of the solar system. This retrograde motion is not understood, and astronomers have been unable to account for it. Venus orbits the sun at a distance of some 60 million miles in 224.7 days.

Mars has also provided us with many surprises in recent years, and almost all the earlier models which have been suggested regarding the Red Planet have been proved completely wrong. This world above all others has been supposed to have been the possible home of life, and even of intelligent life. Gone now is the picture of an old, dessicated world turning into a ruddy desert, with the network of famous canals being a last desperate attempt of the Martians to keep their planet alive by channeling water from the polar caps. The polar caps do exist—they are a very prominent feature seen through a telescope—but they are now reckoned to be a thin layer, perhaps only a few inches deep, of carbon dioxide ice.

The Mariner probes have shown that the famous canals are actually chains of enormous craters, giving an illusion of canals when seen from a great distance. The close-up photographs of the Martian surface taken by the probes have given us our biggest surprises, for they show a surface incredibly ravaged by enormous craters, deep fissures extending for hundreds of miles, and marks which suggest dried-up watercourses or riverbeds. Instrumentation has suggested that the Martian atmosphere is only a thousandth the density at the surface as that of earth, and is mainly carbon dioxide with only minute traces of oxygen or water vapor.

Even so, this tenuous atmosphere appears able to raise vast dust storms which last for many weeks and cover enormous areas of the surface. Mars takes 687 days to orbit the sun at a distance of 141 million miles.

Why these two planets should exhibit such enormous differences between themselves and the earth is a mystery which has never been resolved. Immanuel Velikovsky, in his book *Worlds in Collision,* provides at least one theoretical answer that brought down a storm of protest on his head from the established body of scientific opinion—a storm that still rages.

Velikovsky's theory is as follows:

The planet Jupiter ejected a mass of material that entered the inner regions of our solar system as a comet, traveling in an elliptical orbit that brought it into close proximity with the earth on several occasions. The first was about 1500 B.C., causing the phenomena noted in biblical Exodus—the plagues and the parting of the Red Sea. The gravitational pull of the comet caused upheavals, earthquakes, and tidal waves on a huge and global scale. The comet returned several months later, causing further destruction, and part of its mass rained down upon the earth as fire and sticky vapor. This is how petroleum originated.

After several further near-collisions the comet collided with the planet Mars, interrupting its orbit and causing Mars several times to approach near the earth, causing further destruction. The moon's orbit was interfered with, and its distance and orbit were changed several times. Subsequent to these various encounters the comet lost its tail and became the planet Venus, eventually stabilizing its orbit, and the other planets and moons also gradually settled down into regular paths once more. All these events took place between the third and second millennia B.C.—in fairly recent and certainly historic times.

There are many objections to Velikovsky's theories which we will not dwell on at length here, since an exhaustive study on this subject was undertaken in *Colony: Earth.* The main objection is, of course, that comets are extremely tenuous things: "The nearest one can get to a vacuum and still have a visible object," it has been said. Therefore, a comet of such a nature to have condensed to a planet of the mass of Venus would have had to be so large that it would have covered the entire sky and would possibly have been many times the size of the entire solar system. No phenomenon of such a nature has ever been noted in the annals of ancient times. If there had been it is certain that some astronomers, particularly those of China and Babylonia, would have made a great many notes for the benefit of posterity.

However, Dr. Velikovsky did make some predictions which have proved to be remarkably accurate. Some of his hypotheses regarding the composition of the inner planets contained ideas which were not known at the time (1950), but have been partially verified by subsequent discoveries and space probes.

He predicted a high surface temperature for Venus, which was not too unlikely, considering that Venus is much nearer to the sun than the earth (26 million miles). But he also postulated it had a thick, heat-retaining atmosphere, and that Venus should be rich in hydrocarbons. As we have seen, astronomers have since suggested that Venus may possess seas of hydrocarbons—virtually petroleum oceans—but for different reasons. Velikovsky suggested that Venus had a retrograde axial spin, east to west instead of west to east, unlike the other planets of the solar system. He also said that this rotation would be extremely slow, and so it has proved.

He predicted that Mars, far from being a flattish desert criss-crossed by shallow canals, would be scored and fissured with huge craters and cracks caused by the battering it had received. The planet certainly does bear such marks, as we have seen.

Finally, he said that the moon would be subject to quakes, and that any future astronauts from earth would have to be prepared to experience moonquakes. These do, in fact occur, but they are so slight that they can be detected only by instruments.

It seems fairly certain that Dr. Velikovsky is wrong in several respects. His time scale is in error: the effects reported in Exodus that he ascribes to the actions of the comet are now known to have been caused by the great volcanic eruption of Thera (Santorini) in the eastern Mediterranean. It seems that he may have telescoped a number of events into one, because he appears to equate the Exodus phenomena with legends from other parts of the world which would seem to be more connected with the more ancient Flood mythology. It is obvious that the Flood took place a long time before the Exodus episode.

Also, gravitational forces of the kind he describes would have completely shattered at least one of the planets involved, more likely the smallest, which is Mars.

It has been calculated that the moon, if it were to spiral closer to the earth, could not crash into this planet. When it reached a certain point, known as Roche's Limit, gravitational forces would shatter it. Some of the pieces would fall to earth, causing immense damage, but a greater part of the moon's debris would adopt a new orbit, and we would then have a ring system similar to that possessed by Saturn. It has been suggested that the Saturnian rings may have formed in this manner.

Notwithstanding the objections to Dr. Velikovsky's theory, there remains nevertheless the problem of the peculiarities of the inner planets. It could be that the only logical solution to this problem is that it was brought about by a catastrophe within the solar system. What could the nature of such a catastrophe have been? If it were not the fault of a comet, and we cannot see comets posing such a menace, then we must look elsewhere, and it is possible that the solution to the problem lies in the asteroid belt. The asteroid belt, a huge group of debris between the orbits of Mars and Jupiter, consists of thousands of irregular masses of rocky substance, some of which are almost as large as Mars's tiny moons, and others which are only a few yards across.

On occasion some of these asteroids pass fairly close to the earth. One, called Eros, came within several million miles. The meteors that bombard the earth's atmosphere, some of which are large enough to reach the surface as meteorites, have their origin within the asteroid belt.

There is considerable controversy in scientific circles regarding the origin and nature of the asteroid belt. One school of thought believes that it is the remains of the material left over from the original gas and dust cloud when it condensed into the solar system. This mass forever orbits the sun as a loose conglomeration of particles, unable to form a planetary body because of the proximity of Jupiter's strong gravitational field. However, many of the asteroids are irregular in shape, and it may be thought that even though a planetary mass could not form, the particles, being dust and gas, would form small masses that would be regular in shape. It would therefore be more likely that to fit this hypothesis the asteroid belt should consist of numberless small spheres orbiting the sun.

The other body of opinion states that the asteroid belt is the remains of a planet that used to orbit between Mars and Jupiter and was for some reason shattered. This event is placed as having happened many millions of years ago, possibly during earth's mesozoic period.

The reason for the dispute centers around those visitors from the asteroid belt which have from time to time landed on earth—the meteorites. Several of these have been of such a composition as to suggest that they were formed within a planetary body of some considerable size.

For example, there is the Orgueil meteorite, which caused a great deal of controversy and in fact still does. This is said to contain "organized elements" which, under a microscope, resemble fossilized microscopic life forms.

Other meteorites have been found to contain diamonds, which can only be formed naturally under great heat and pressure, again pointing to their formation within a planetary body.

A fragment of a coral substance fell from the sky and was collected by an American, Donald Bunce. If it were terrestrial coral, it could have fallen from an aircraft. If not, did it perhaps come from the asteroid belt, encased within a mass of stone which disintegrated on its path through the atmosphere, leaving the coral intact? Coral, we know, is a product of marine animals, and *if* this came from the asteroid belt, it points not only to a planet destroyed, but to a planet which had once borne life and, moreover, a world where the temperatures must have been high enough to have supported tropical or subtropical forms of life. It would be one of a piece with the meteorite containing the mysterious fossilized microorganisms.

If a planet had orbited between Mars and Jupiter, what could have shattered it? Astonomers consider that there are no known *natural* forces which could have disrupted a planet of any size.

There is one possible *natural* catastrophe that could account for the asteroids: if an intruder body of considerable size came in close proximity to the hypothetical planet, gravitational forces of the kind already described would come into play and the planet would be torn apart. As there are no traces of such an intrusive body within this system we would have to assume that both bodies were shattered by gravitational stress.

There are two objections to this particular theory. One is that it appears extremely unlikely that there are bodies in space just floating between the stars—all stellar objects we have detected would seem to be in association with each other because of gravitational attraction. As the galaxy as a whole rotates in a whirlpool fashion, it is extremely unlikely that any stellar object would be able to deviate from the galactic spin.

We have mentioned this factor in connection with the hypothesis of a wandering star colliding with the sun, thus causing the creation of the planetary system. Although many novae and supernovae have been observed during the past several thousand years, we have never noted colliding stars or planets—although colliding *galaxies* are not unknown. It would seem to be even less likely in a thinly populated region of space, such as the sun occupies, than in the more central regions, where the stars are clustered closely together.

Secondly, if the asteroid belt were the remains of *two* planets, or even a planet and a large moon, it is doubtful if the mass of either of them would be large enough to support life or produce diamonds from its interior.

However, the idea that the asteroid belt is the remains of a planet orbiting between Mars and Jupiter has some merit, and it would be one answer to some of the peculiarities we find within the inner planets of the solar system.

Let us imagine that such a hypothetical planet had been destroyed in the position occupied by the asteroid belt. The nearest other world to this catastrophic event is Mars, which could be expected to bear the full brunt of the debris scattered in all directions from the explosion. We could therefore expect the surface of Mars to be severely scarred. The impact of masses of rocky debris would produce huge craters, larger than any we see on the moon, and the resulting gravitational disturbance could cause a disturbance in Mars's orbit, which, although slight, would have a devastating effect, causing vast earthquakes on the Martian surface. Possibly, the Red Planet may have come close to destruction itself.

What do we find on Mars? We find a planet hideously scarred by enormous craters, many of them larger than the lunar craters, and there are great cracks and fissures extending for hundreds of miles.

Next in distance from Mars we have the earth-moon system. Both of these bodies would experience a rain of debris. The moon, being airless, would have less shielding from these missiles than the earth, and as a consequence would be heavily cratered. This, of course, is what we find.

There is one aspect about the moon that does seem somewhat odd. If, as many astronomers believe, the craters on the moon are impact craters, why is it that no recent craters have appeared? We know, of course, that over many millions of years the moon could gradually have taken on its present appearance by a slow and continuous bombardment from the asteroid belt; but surely, in the hundreds of years that the moon has been observed, at least one impact of considerable size should have been seen. The earth is subjected to a ceaseless bombardment of meteoric material, and some must have been of considerable size to reach the earth's surface in large chunks—remember, most of a meteor's mass is vaporized by passage through our dense atmospheric envelope. In the case of the moon, there being no atmosphere, one should expect many sizable bolides to strike the lunar surface, gouging out craters at least large enough to be seen through our powerful telescopes.

Also, it never has been pointed out that a meteoric object orbiting in space in the normal way, approaching the earth-moon system, would surely be most likely to strike the earth, since the gravitational attraction of the earth is so much greater than that of the moon. If the enormous number of craters on the moon are due to meteoric impact, then there should be evidence of many more on earth than on our satellite. They would be less easy to detect today, of course, because erosion by climate, earthquakes and vegetation would almost have obliterated them.

At the present time there are on earth some twelve large craters that are thought to be caused by meteoric collisions. There are a further thirty-two that are considered *possibly* to be the remains of impact craters. The largest of the meteoric craters is the Barringer crater in Arizona, which is some 4,200 feet in diameter and 570 feet deep. The rim is some 160 feet above the floor of the plain. No traces of a large meteorite have been found in the vicinity of the crater, and digging below the floor of the crater has failed to produce any remains. Prior to 1902 numerous nickel-iron specimens were collected from the vicinity of the crater, and this tends to support the impact theory, since many meteorites are composed of nickel-iron.

It may seem reasonable to assume that the explosion of a planet orbiting beyond Mars would account for the cratering of Mars, the moon, and, to a lesser extent, the earth. The fragments flung from the explosion toward the far regions of the solar system would scarcely affect the outer planets, the nearest, of course, being giant Jupiter, which is thought to be more gaseous than solid and would therefore sustain no damage. In any case, even if solid, its great mass would not be affected by such impacts.

As far as can be calculated, the present mass of all the debris in the asteroid belt is some 1/1,000th of the earth's mass. To judge from the vast number of craters on Mars and the earth/moon system, and taking into account also the many meteors and the meteoric dust assimilated by earth alone over the centuries, it is not impossible that a mass of planetary dimensions, albeit smaller than earth and Venus, once existed in the region now occupied by the asteroid belt.

As we have said, many astronomers do not agree that there ever was a planet orbiting between Mars and Jupiter, but there is a measure of agreement that the moon's craters appear to have been formed at one particular point in time, which is why there exist two schools of thought as to their formation. One school favors the impact theory, whereas the other point of view holds that the craters were formed by volcanic action in the moon's early stage of evolution. However, the extensive cratering of Mars, where there exists the possibility that the craters were formed long after the planet had ceased to evolve, and the fact that craters similar to those of the moon exist on earth in areas where there is no trace of vulcanism, would tend to suggest that the impact theory is more likely.

If, then, the craters on the moon, Mars, and earth were formed at the same time, there must have been a prior cause, an event of tremendous violence. This could have been the explosion of a planetary mass orbiting between Mars and Jupiter. If this were so, it must have been an explosion of exceptional violence for the fragments to have been flung at great velocities across the millions of miles that separate the asteroid belt from Mars and the earth/moon system.

Astronomers have said that they know of no natural event that would lead to the destruction of a planetary mass except the interference of another planetary body. Such an event is possible when two bodies come into close proximity with each other, where gravitational forces will cause disintegration of one or both of the bodies; but we have seen that there was only enough mass to make up one planet. Therefore, if there was only one planetary body involved, was the cause of the explosion artificial?

This would imply the activities of an intelligent race, and here we have several possibilities to choose from. The weapons that could cause an entire planet to explode are unknown to us but could possibly have been nuclear devices of immense power. Weapons based upon the control of gravitational fields would be capable of causing such destruction. We have mentioned in Colony: Earth that control of gravity would be not only a tool of tremendous value but also a weapon of awesome power.

Was our hypothetical planet inhabited? If it could be proved that the meteorites do contain fossil material, and that they do originate in the asteroid belt, then it is more than probable that they could be the remnants of a destroyed planet which was life-bearing—it would not be too far from the sun to be so. And where there is life, there is also the possibility of intelligent life.

Was there more than one inhabited planet in this solar system at some point in the past? We have at least three possibilities. Earth we know to be inhabited. There is the possibility based on the evidence of the meteorites that there was a life-bearing planet orbiting beyond Mars, and there is Mars itself.

The Mars photographs taken by the Mariner space probes have revealed what looks like the beds of dried-up watercourses. They have the same meandering appearance, with many tributory branches. The probes have also revealed that the Martian atmosphere is extremely tenuous, only 1 per cent of earth's at surface level, and appears to contain no free oxygen. This has surprised our scientists, being much thinner than was thought, and the lack of atmosphere is difficult to account for. It has been suggested that in the past Mars may have had a considerable atmosphere, so that if some of the surface markings really are the remains of watercourses then there must have been a considerable amount of surface water and presumably an atmosphere not too dissimilar from earth's.

What, then, has happened to the Martian atmosphere? One possible solution suggested by the scientists is connected with the fact that the Mariner spacecraft were unable to detect an appreciable magnetic field on Mars, which could suggest that the original iron content of the planet did not sink toward the center, as in the case of earth, but remained on the surface. Water vapor in the atmosphere is separated by the action of ultraviolet light into oxygen and hydrogen. Hydrogen, the lighter gas, could escape into space, and the oxygen would then combine with the iron, forming a rust compound. This theory would account for the ruddy appearance of Mars. However, this is only a tentative answer to the disappearance of the Martian atmosphere, and its elimination may have taken a more violent form.

Consider this possibility: If we have three inhabited planets within the solar system and a conflict arose among the intelligent inhabitants, possibly one planet was completely destroyed and perhaps Mars was so severely damaged by the debris and the use of highly destructive weapons that its atmosphere was vaporized. Perhaps the "war in heaven" mentioned in the ancient annals was a conflict between warring worlds.

We have suggested that man may be of extraterrestrial origin, and have postulated that vehicles from other solar systems once surveyed this system and found not one but three planets suitable for life, or perhaps already bearing life of some kind. Perhaps both Mars and earth, and possibly Venus, were occupied by these visitors, and the outer, now-destroyed planet was originally used as a defensive installation. We are considering using the moon for human occupation; we may also consider building missile installations there and using the moon as an impregnable military base that could threaten any part of the earth. It is possible that something of the sort was carried out in the remote past.

We have considered the possibility of a nuclear conflict on earth in the distant past; we should perhaps also consider whether this possibility was limited to earth or was carried to other worlds.

A past civilization that had developed nuclear weapons and missiles would doubtless also have developed space travel, even if only of a rudimentary kind; this is the stage at which we have now arrived. If they had made their discoveries in a different order they might have perfected space travel first.

It came as a complete surprise to the astronauts and to our scientists, but it appears that the moon is covered with "glass." A vast quantity of small glass spheres were found on the surface of the moon, and it has been suggested that these were formed by the high temperatures obtained by impact with high-velocity meteorites. However, the temperatures needed to vitrify rock to this extent are extremely high, as we have seen. But would meteorites striking the moon's surface do so with the hypervelocities necessary?

The gravitational attraction of the moon is only one fifth that of the earth, and we have no terrestrial evidence to show that meteoric impact causes this particular phenomenon. With our gravitational field one would expect the velocities to be correspondingly higher, notwithstanding the cushioning effect of our atmosphere. The frictional effect of the passage through atmosphere, raising the temperature of the meteorite high enough to melt a considerable proportion of it, should itself create much higher temperatures.

On the face of it, it does seem more likely that a cold meteorite striking cold surface rocks—there being no atmosphere on the moon either to impede its passage or to raise its temperature by friction—would merely cause pulverization. Indeed, it has been thought that the layer of fine dust which covers the surface of the moon is caused partially by the eroding away of the surface material by the great contrast in temperatures between night and day and partially by the pulverization of meteoric material. If this pulverization is one of the reasons for the moon's dusty layer, then it can hardly be responsible for the glassy spheres, for which extremely high temperatures are required.

There are also lunar areas of glasslike material which remind us of the vitrified areas that exist on earth, particularly in present-day desert areas, and which show a marked similarity to vitrification caused by recent nuclear tests in desert regions. Could the similar phenomena on the moon have had a similar cause—thermonuclear heat?

I quote here from *The Solar System* by Frank W. Cousins (John Baker, 1972), Chapter 10, "On the Earth/Moon System," page 177:

"The discovery that some parts of the Moon are paved with pieces of glass supports the view that the Moon has suffered impacts of a very energetic nature. The report of these findings in 'The Times' of September 2, 1969, drew the following letter from Mr. D. O'Brien of Gonville and Caius College, Cambridge:

" 'Sir: It is satisfying to read your report today that the Moon is made largely of glass. This is just what Empedocles in the fifth century before Christ said it was made of.' "

From where did this Greek obtain his truly remarkable information? Was it guesswork, an assumption it would need a glassy surface to make it shine so brightly by reflected sunlight? Did he know that the moon shone by reflected light? Perhaps he had access to more ancient *knowledge*. We shall, at a later stage, examine some areas of astronomical knowledge from ancient times which does not seem possible to have been obtained by naked-eye observation alone.

Considering the lunar glass being formed by temperatures in the range created by thermonuclear reactions, we turn our attention to those mysterious objects found in certain areas of the earth which are reckoned to have an extraterrestrial origin, including the suggestion of a lunar origin: the objects known as tektites.

Tektites are a silica-rich obsidian glass different from terrestrial obsidian. The greatest number have been found in Australia, Indochina, the Philippines, and Moldavia.

They are usually jet-black and take the form of button shapes, spheres, and dumbbells. Some consider that they are extraterrestrial in origin and have gained their shapes by aerodynamic ablation through high-speed flight and kinetic heating on their passage through the earth's atmosphere.

Chapman *(Nature* 188: 333, 1960) is of the opinion that they originated on the moon, but this view is not universally accepted and their origin is still uncertain.

Urey has suggested that tektites may have originated in a cometary collision with the earth. He therefore suggests that they are terrestrial in nature, produced by extremely high temperatures. He likens the explosive heat to that produced by a high-altitude burst of a hydrogen bomb. It is, in other words, a "blast burn" effect.

If the effect that produced tektites is almost or completely identical to a nuclear explosion, then perhaps this is how they were in fact produced. We have a certain amount of evidence to suggest that nuclear weapons were used on earth in the past, although modern nuclear weapons have failed to produce tektites. Did they result from a nuclear explosion either on the moon or from similar explosions that may have been responsible for the destruction of a planet orbiting beyond Mars? The difference between the vitrified areas on earth and the moon and the moon "marbles" may lie largely in the fact that these particular objects were subjected to high-velocity flight through the earth's atmosphere.

It is not impossible that tektites have the same origin as meteorites and represent fragments of the vitrified areas of a planet shattered by nuclear explosions of great power.

The destruction of a planet beyond Mars could thus furnish us with an explanation for the appearance of Mars and the moon, and for some of the craters on earth. When this event took place is difficult to say. Most astronomers who favor the theory that the asteroid belt is the remains of a planet locate the event many millions of years in the past, but this may not necessarily be so. The event could have happened only some thousands of years ago and be related to the legendary war in heaven. It is interesting to observe here part of the legend regarding the reason for the construction of the Great Pyramid:

"King Saurid, son of Salahoc, reigned in Egypt three hundred years before the Flood and dreamed one night that the earth was convulsed; all the houses fell down upon men and the stars collided in the heavens such that their pieces covered the sun. The king awoke in terror, rushed into the Sun Temple, and consulted the priests and diviners. Akliman, the wisest of them, said he too had had a similar dream ... it was then that the king had the pyramids built in that angular way suitable for withstanding *even the blows from stars. . . .*" (Italics mine.)

Blows from the stars! Could this be the event we have described as possible: the destruction of a planet and the bombardment of its fragments? If so, this would place the event within the last ten thousand years. We can date the age of the rocks that compose asteroidal meteorites, but we cannot so easily date the time when they came to be in their present fragmentary state.

This, of course, does not solve the riddle of the other inner planet of the solar system, namely Venus. It may be that Venus orbits too far from the center of the explosion to be affected, provided it was then following its present orbit.

Velikovsky, in *Worlds in Collision* says that at one time Venus did not occupy the prominent place in the sky it occupies today, that this position in the eyes of the ancients was occupied by Jupiter. Venus is such a bright object it seems surprising that it may not have occupied a prominent place in ancient astronomy after the sun and the moon, for it is far brighter than Jupiter. Either there was a time when Venus did not possess its dense atmosphere, and therefore lacked its high albido, or it orbited much farther from the sun.

We see the solar system as it has been known for the past several thousand years, and therefore our astronomers have a firm opinion that this is how it has been for countless millions of years. This view may not be correct. There could have been great changes in the system due to catastrophic events of which we are not aware and which have left no observational traces now. Because the system has been stable for the six thousand years for which we possess written records, it should not be automatically assumed that this is how things have always been.

It is possible that when we are able to undertake human exploration of the inner planets we will be able to answer some of these riddles. Already our first tentative explorations of the moon have provided many surprises, and some features of our satellite have been startlingly at variance with what we have thought from our previous observations.

Possibly the surprises that await us on the surface of Mars and Venus will be even greater than those we have encountered on the moon.

7

Ancient Astronomy

A REEXAMINATION of ancient scientific thought and writings leads to the assumption that science has a much longer history than has been thought, and that much of what we know today is not a discovery but a rediscovery of ancient and half-forgotten knowledge. It appears also that the more ancient the source, the more basic and profound the knowledge that existed. This, like the origin of civilization itself, is the reverse of the situation that should exist if the first dawn of civilization was at Egypt and Sumer.

Knowledge existing in antiquity regarding astronomy, for one important example, provides us with many surprises. Most authorities give us the impression that early astronomy was limited to those objects visible only to the naked eye, and was mostly falsely interpreted and connected intimately with superstition and astrology. This is true enough of many aspects of later Babylonian, Egyptian, and early Greek thought, which show us a simplified and often totally erroneous picture of the universe. A flat earth; the sky a solid dome with holes in it appearing as stars; the earth held up by a turtle swimming in the primeval ocean; the earth riding on the back of an elephant; Atlas holding up the sky, and many other fanciful ideas which seem absurd today.

Even in the Middle Ages knowledge of the universe was extremely limited, and the idea that the earth revolved around the sun was held to be ridiculous—the earth was fixed at the center of the universe, and everything revolved around it.

For the past several hundred years everyone has known that the earth is a sphere (although there is still, even today, a small Flat Earth Society), but throughout medieval Europe everyone was taught that the earth was flat. It was heresy, and rather dangerous, to think otherwise. Even today, in an advanced country like England, it is surprising how little known astronomical truths are to the majority of people. Many people, for example, thought that the first lunar astronauts would somehow fall off if they weren't careful, and others feared this sort of experimentation would interfere with the terrestrial weather. I personally have heard such comments. This, in an enlightened country in the latter part of the twentieth century!

It is hardly surprising, therefore, that earlier peoples from nontechnological societies would have an even more difficult time grasping astronomical realities, and even those literate and educated folk of the time would not comprehend them. Little wonder that the knowledge stemming from the remote past was either not understood or ignored or dismissed by the then leaders of thought, scientific and religious.

Thousands of years ago it was known that the earth was a sphere and hung, unsupported, in space. This fact is even mentioned in the Bible: "The earth hangs in nothingness," it says—a point that appears to have been ignored for most of the lifetime of Christianity. It seems that even Christian societies believed what they wanted to believe and ignored the rest.

Many ancient thinkers were aware of the true shape of the earth. "The earth is round and it revolves around the sun," said Anaximander (610–547 B.C.). Pythagoras said in the sixth century B.C. that the earth is a globe.

King Chandragupta told the Greek ambassador Megasthenes in 302 B.C.: "Our Brahmins believe the earth to be a sphere."

Many ancient thinkers not only knew that the earth was a globe, but they estimated its size, its orbit and axial rotation.

Aristarchus of the third century B.C. said: "The earth revolves in an oblique circle while it rotates at the same time about its own axis."

"The Earth spins on its axis once in twenty-four hours." So said Heraclides of Pontus in the fourth century B.C..

Eratosthenes measured both the circumference of the earth and its diameter. There is a discrepancy of only about 110 miles between the figure he obtained for the polar diameter and that shown by modern astronomers.

The Sanskrit book *Surya Siddhanta* contains fairly accurate calculations of the diameter of the earth and its distance from the moon.

Chang Heng of China (A.D. 78–139) said the earth is an egg, and that its axis pointed to the Pole Star.

In more modern times Columbus made a study of all available classical sources before embarking on his voyage of discovery to the New World. In a letter preserved in Madrid he made the remarkable statement that the earth was slightly *pear-shaped*. It is only in the last ten years that space satellites have confirmed this fact, one previously quite unknown to our astronomers. From what ancient lost text did he discover this information?

Parmenides of the sixth century B.C. says, about the moon: "It illuminates the night with borrowed light"—an obvious reference to the fact that the moon is illuminated by reflected sunlight.

Empedocles (494–434 B.C.): "The moon circles around the earth—a borrowed light."

We have already mentioned the Sanskrit legend about the lunar Pitris and the great age of the moon, and it is not only in India that the moon was regarded as being older than the earth. In Mayan art the moon is represented as an old man with a conch shell. The moon goddess Ixchel of Mexico was known as the Grandmother. In the religions of many primitive peoples the moon is considered to be the first man who died, the Encyclopaedia Britannica says.

It is only since our astronauts landed on the moon that we have learned that the moon is indeed older than the earth, and that its composition is different from this planet's.

The most extraordinary astronomical calculations were made

by the astronomers of the Mayan civilization of Central America. The Copan astronomers estimated the lunar month as 29.53020 days, and the Palanque astronomers as 29.53086. According to our astronomers the period is 29.53059 days, midway between the Palanque and Copan calculations of the ancient Mayas.

Moving away from the earth/moon system, even more extraordinary facts have come to light about ancient astronomy. There are Babylonian inscriptions that mention the Horns of Ishtar (Venus), which refer to the crescent shape of the planet. The "horns" of Venus can only be seen through a telescope.

Babylonian priests recorded their observations of the four greater satellites of Jupiter—and they cannot be seen with the naked eye. Professor G. Rawlinson says of them: "There is said to be distinct evidence that they observed the four satellites of Jupiter and strong reason for belief that they were acquainted likewise with the seven satellites of Saturn."

The Dogons of the Sudan have a curious legend about the "dark companion of Sirius." The dim companion star of Sirius can only be seen through the most powerful of our present telescopes, such as the two-hundred-inch mirror of Palomar. Also the Dogons say there are three stars, one of which, brighter than iron, is so heavy that a tiny grain would weigh more than 480 donkeyloads. We have calculated that Sirius B's density is fifty times that of water and that a small box of grains would weigh a ton. There also is a suspected third star.

Democritus of Greece, in the fifth century B.C., said: "Space is filled with myriads of stars, and the Milky Way is but a vast conglomeration of distant suns."

Aristarchus said that the distance which separates us from the stars is immeasurable.

How did these early Greeks know this? It is curious that the ancients spoke of looking at the sky and distant objects "through tubes." Telescopes in antiquity?

Heraclitus (540–475 B.C.) thought that each star was the center of a planetary system. Democritus said that other worlds come into being and die, and that only some of them are suitable for life. Anaxagoras (500–428 B.C.) wrote, "Other earths produce the necessary sustenance for inhabitants." The Vedas of India speak of life on other celestial bodies far from the earth.

"Comets move in orbits like the planets," writes Seneca. Aristotle cited the Pythagoreans, who taught that comets were stellar bodies that appeared and reappeared after long periods of time.

Our present idea of the formation of the solar system out of a flattened disc of gas and dust was anticipated in ancient times. The Maya *Popol Vuh* says: "Like the mist, like a cloud, and like a cloud of dust was the creation."

The *Huai Nan T'zu* and the *Lun Heng* written by Wang Chung (A.D. 82) stated that worlds were made out of whirlpools of primary matter.

We are all familiar with the signs of the zodiac, and the names which have been given to the constellations: Orion, Taurus, Aquarius, Pisces, etc. However, it takes a good stretch of the imagination to see these stellar groups taking on the *appearance* of the descriptions that have been given to them: an Ox, Hunter, Water Carrier, and so on. It seems all the more surprising, therefore, that many different cultures from all over the world have the same names for the constellations. One might have thought that various cultures would have interpreted these star groups differently and invented descriptive or fanciful names very different from each other. Such is not the case.

Orion, the Hunter of the Middle East, is known as the Hunter of the Autumn Hunt in China.

The Western Aquarius becomes Tlaloc, the Rain God of Mexico—both connected with water.

The Babylonian sign of the Ram is the Sheep in China.

The Chinese Ox constellation is Taurus (the Bull) in the West.

Giorgio de Santillana says in *The Origins of Scientific Thought*: "They [the names of constellations] were repeated without question substantially the same from Mexico to Africa and Polynesia—and have remained with us to this day."

Tomas, in his book *We Are Not the First*, has suggested that early civilizations may have had access to older lists of constellations that they used to identify the stars.

All these things we have mentioned would seem to point to there having been a greater knowledge of observational astronomy in the past than has usually been realized. There may also have been, to judge from the Dogons and their mention of Sirius, a knowledge of astrophysics, most of which has been lost.

Some of the knowledge that existed could not have been obtained without instrumentation. The moons of Jupiter and Saturn cannot be seen without the use of telescopes, and it goes without saying that the dim companions of Sirius definitely are invisible except through a very powerful and sophisticated telescope. Yet we have not been able to find traces of telescopes from the remote past, although small lenses capable of slight magnification have been found in Mesopotamia, which shows that a knowledge of optics did exist in the past.

How did the Maya arrive at their accurate computations regarding the lunar month? Such minute fractions of a day cannot be obtained without precise and accurate instrumentation. As far as we are aware, the Maya possessed no such instruments: this civilization hardly ever even used metals. There is what appears to be an astronomical observation at Chichen Itza in Yucatán—the shape is almost identical to modern observatories, minus the scientific equipment. Did such a place once hold astronomical devices? Or were the Mayas merely copying a structure from an earlier, vanished era, knowing its connection with the heavens, but either unaware of the equipment it should house or unable to construct it so that the building could fulfill its true function?

Perhaps they were carrying out a sacred task, in the same way that the Brahmins faithfully copied the mathematics of a vanished science, because it was a sacred duty, and therefore an act of homage to the gods. Did the Mayas actually make these calculations of the lunar month, or did they, as with the observatory, have this information transmitted down to them from a vanished scientific era? This again could be a close parallel to the Brahmins.

We can suggest that a certain amount of knowledge of observational astronomy has survived from an ancient period, some of which can only have been obtained with the aid of advanced instrumentation. This presupposes a legacy from an advanced scientific era of which no trace remains. If such an advanced society possessed the devices of astronomical instrumentation, then there should also have existed some knowledge not only of observational astronomy, but also of astrophysics.

Very little evidence does in fact exist, but there is enough to make us pause. The ancients were aware of some things outside of their earthbound state that, properly speaking, they should not have known unless someone, at some time in the past, had actually experienced them. This is not to say that an ancient Sumerian or Babylonian had experience of space personally, but they may have had traditions or documentary evidence to draw upon which have survived in their writings.

The Babylonian *Epic of Etana,* written 4,700 years ago, has drawn comments in recent times because it presents a picture of the earth seen from a great height, possibly from space, which is so accurate it could not have been based on guesswork:

" 'I will take you to the throne of Anu,' said the eagle. They had soared for an hour and then the eagle said: 'Look down, what has become of earth!' Etana looked down and saw that the earth had become like a hill and the sea like a well. And so they flew for another hour, and once again Etana looked down; the earth was now like a grinding stone and the sea like a pot. After the third hour the earth was only a speck of dust, and the sea no longer seen."

Anu, the Great God of the Babylonians, was god of the Heavenly Great Depths—which we today call space.

If the earth was what the later ages of Babylon thought—the

center of the universe around which all the stars were fixed—a concept that the earth could shrink through distance to the size of a speck could not have been imagined. More ancient times had a more accurate view of the earth than that.

The infinite reaches of space are accurately described in the Egyptian *Book of the Dead:* "This place has no air, its depth is unfathomable, and it is as black as the blackest night."

An ancient legend recounts that in the time of the Chinese Emperor Yao a great disaster overtook the earth; the seas rose and covered all the land, there were plagues of fire and great earthquakes. This emperor had an engineer called Chih-Chiang-Tzu-Yu who voyaged to the moon on a celestial bird. This bird knew the path, and traveled by mounting currents of luminous air. Once in space, the Chinese astronaut "did not perceive the rotary movement of the sun." On the moon, he saw the "frozen-looking horizon." It appears that on one occasion his wife Chang Ngo flew to the moon—"a luminous sphere, shining like glass, of enormous size and very cold."

There are some statements in this passage which take the story out of the realm of fairy tale. The celestial bird gave him directions—was this information from a computer, similar to that obtained by present-day astronauts? The currents of luminous air could refer to the fiery jets of a rocket propulsion motor. Again, we have a parallel to the ancient Greek statement about the glassy appearance of the moon, which we have found to be true only *since men set foot on the moon.*

An ancient Sanskrit text, the *Surya Siddhanta*, makes mention of Siddhas and Vidyaharas (philosophers and scientists) who examined the earth "below the moon but above the clouds." This could relate to scientific surveys undertaken in something like our present-day orbiting space laboratories. This is an extremely interesting point, because of a feature to be seen on the map of Piri Ris, which shows the Antarctic and South American continents and is said to date from before the time of Alexander the Great. There is a distortion on this map which exactly fits the shape of South America when seen from a satellite orbiting *at a height of some seventy miles.* Was it from an orbiting space station of the type described in the Sanskrit text that the original of this map was made?

The *Samaranagana Sutradhara,* also from India, speaks of a time when men flew in the air and *heavenly beings* came down from the sky.

Professor H. L. Hariyappa, of Mysore University, wrote that in a remote period "gods came to the earth often, and some men visited the immortals in heaven."

The *Mahabharata*—the now-famous text which mentions nuclear weapons—says in its fifth volume: "Infinite is the space populated by the perfect ones and gods, there is no limit to their delightful abodes."

In an ancient Chinese book, the *Shi Ching,* the Divine Emperor saw the crime and vice arising on the earth (do we not have echoes of Genesis here?) and commanded Chong and Li to cease communication between the earth and sky "and since then there has been no more going up and down."

In the Royal Pedigree of Tibetan Kings, it is said that the first seven kings came from the stars and were able to "walk in the sky."

What are we to make of all these extraordinary statements connected with astronomy and cosmology in ancient texts? Some indicate astronomical knowledge of an advanced type unobtainable without instrumentation; other information could only have been obtained by flights into space, and there are references to actual travel in space, both by terrestrials and apparently by visitors from the stars.

Consider the observational astronomical knowledge of the Greeks and Babylonians. Where did they obtain their knowledge in the first instance? Some of the statements by the Greek thinkers may have been educated guesses, but on the other hand, since they are invariably so accurate, had they read them elsewhere?

The cradle of knowledge of ancient Greece, and this may also be true of Babylonia, was Egypt. Greek thinkers and students went in unknown numbers to the wisdom of Egypt for their knowledge. There was a great university in Alexandria with a library of 700,000 books in scroll form, supposed to contain all the knowledge and history of the human race. The Bruchion contained 400,000 books and the Serapeum 300,000. The university also included facilities for the study of medicine, mathematics, astronomy, botany, and zoology, and it could house 14,000 students. The university and its great library were razed during Julius Caesar's Egyptian campaign.

Was this where the sages of Greece obtained some of their knowedge of astronomy? But if so, where did the Egyptian priests and men of wisdom obtain all these books and all the knowledge they contained?

We perhaps have a clue from legend about the true function and purpose of the Great Pyramids. In the myth, King Saurid, whom we have already mentioned, was warned in a dream that there was to be a great flood, and he was advised to build a great pyramid to store *all the world's knowledge.* According to this, the pyramids were not built as tombs, but as shelters, not for people but for knowledge.

It is known that the pyramids were built by a technique that cannot be duplicated today. For all our technological achievements we would be unable to build the Great Pyramid of Cheops—we simply lack the capabilities. If we assume the builders to be the representatives of an advanced civilization aware of a great catastrophe about to overtake them, then it is possible that they secreted this knowledge of the sciences within the pyramid, to be brought forth when the danger was past. It is this knowledge that was the basis of the Alexandrian library.

We shall never know the extent of the loss to humanity caused by the destruction of this library. If the knowledge handed down to us by the Greeks is but a minute fraction of what they remembered or recorded from their studies in Egypt, the loss is probably incalculable. Possibly there was more knowledge of the sciences contained in this one library alone than what we have so painfully gathered during the ensuing two thousand years. And the Alexandrian library was not the only one in Egypt—the documents in the library of the Temple of Ptah at Memphis also were totally destroyed.

The surviving fragments of knowledge from the past in many fields—astronomy, medicine, botany, engineering—demonstrate that there must have been an advanced civilization in the past, as much of this knowledge could only have been gained by the use of sophisticated technology.

As we have seen, there are some things it seems impossible to discover except by actual exploration, such as the knowledge that glass exists on the moon. We have discovered this fact by going there, and it now seems within the realm of possibility that our recent space flights may not be the first from this planet. Have men from a forgotten civilization trodden before on the surface of the moon?

The ancients mention that other stars have planets; how could they possibly have thought this unless such a fact had been known at some time in the past? It is only within the last several decades that our astronomers have come to the conclusion that possibly all solar-type stars possess planetary systems. It is now thought that there must be millions of planetary systems in the galaxy, and therefore thousands of earth-type worlds. This is just what the ancients said; statements dismissed as fairy tales are now being verified by our scientists!

How much else of what the ancients have written is true? Let us examine the statements that men visited other worlds in the past and that heavenly beings—these must be the inhabitants of other planets—came to earth. The old legends clearly indicate that men understood the nature of the space that exists between worlds. This knowledge could best be gained by experiencing the reality of these gulfs, and the writings do indeed seem to be based on these experiences rather than theoretical speculation.

A highly advanced civilization of the past, capable of space travel and perhaps travel to the stars, may have had contact with other cultures in space. The theories we have advanced—that mankind on earth may have been an offshoot of a galactic civilization—would mean that such contacts were feasible.

It is also possible that there existed channels of communication between the civilization here and civilizations elsewhere in the galaxy. It could be wondered: Is this communication with advanced cultures in space—in the heavens—the true origin of prayer? All through history men have prayed to the gods, who dwelt—invariably—in the sky. The ancients talked knowingly of other worlds in space, and there are legends of the visits of the heavenly ones to earth. Prayer, then, may be based originally on communication between the inhabitants of this planet and a galactic civilization.

Perhaps the communication was on a telepathic level; we have evidence to show that telepathy is possible. If telepathy is one of the extrasensory powers that now lie largely dormant in the human mind, we cannot be sure that this was always so. There may have been a time when these powers were active, and they could have atrophied with the passage of time. If it were true that man is a degenerate species, this degeneration would also affect the mind.

We also do not know whether telepathy would be subject to the limitations of electromagnetic radiation. It is one method of communication which, unlike any form of radio or television transmission, may not be subject to the limitations of the velocity of light, and may therefore make contact virtually instantaneously with other beings many light-years away.

As the gods are invariably situated in the sky, and the sky is where other worlds exist which may contain intelligent forms of life, then the gods must be the inhabitants of those other worlds. Telepathic communication may be one method of passing information—and perhaps also orders—between man on earth and the "gods" in space. But there may be other methods . . .

It is interesting in this connection to take into account two factors from mythology that have a bearing on this subject. One is the curious statement which occurs time and time again that "time" itself is different for mortals and for the gods, and indeed for human beings when they traverse the heavenly gulfs. We have already noted that a day of Brahma is a vastly longer period of time than a terrestrial day, and that the same thing applies to the Egyptian Horus. We also find this in the Judaic-Christian tradition, where it is said of God "that a thousand years are but a moment in Thy sight."

The vision of Isaiah contains a curious story which has a connection with time differences away from the earth. He was taken to heaven by an angel to see God. The angel then told him it was time to return to earth, and Isaiah, surprised, said, "I have been here but two hours." To which the angel replied, "Not two hours but thirty-two years." Isaiah was grieved to think that such a long journey would mean old age or death, but the angel told him that he would not have aged on his return.

In Exodus, we remember, Moses went up to the Mount to meet God. A strange cloud came down with smoke, flame, and thunderings, and a loud voice told the people to stand back lest they be killed. Then Moses saw the feet of God standing on a pavement as if of crystal, and was warned not to look at the face of God.

In modern terms, how could we translate this passage? Was the strange cloud a large space vehicle, perhaps windowless? If it was highly reflective (our lunar vehicles are covered with a reflective layer to prevent excessive heat absorption), then it would mirror the clouds and sky and may well appear to the natives to have taken on the appearance of a cloud itself. The flame and thunder could be the exhaust from braking rockets, fatal to approach, as anyone who has observed the ascent of a Saturn rocket would realize.

The loud voice could have been broadcast from a loudspeaker. Seeing the feet of God means that what Moses saw was a *person*. It was no amorphous mental image he saw, but an actual personage, standing on, perhaps, a platform extended from the ship. Why was he not allowed to see the face of God? Was it perhaps that "God" was wearing a helmet with all the attachments of an astronaut, with a visor hiding his features, and did not wish to terrify the primitive any more than necessary?

This concept of a god who occupies a certain limited physical space is entirely different from the view of God possessed both by modern thinkers and by the basic religious concepts of deism, and has more in common with some aspects of modern scientific thought than with primitive superstition.

The "God" who has feet and stands is a person, whether or not this person has an extended life span that may make him seem immortal and has powers at his command that would seem awesome even to us today. Such an entity could not, by our standards, be regarded as "God" as we understand the term, and could at least be comprehended by modern scientists, even if not properly understood.

In this connection, we may note that the Mormon sect of Christianity (the Church of Latter-Day Saints) has as a main tenet of its faith that God is a person of "flesh and bones," as they put it. They even believe in the concept that this God has a planetary home in the galaxy. This idea, which almost seems science fiction, is proposed quite seriously, and certainly echoes more ancient concepts of the "gods from the sky."

The other curious aspect of this God of the Hebrews, in this context, is that he would seem to be aware of events which had happened many thousands of years before. He apparently knew what their ancestors had done. This means either that the God had a life span of many thousand of years, or that he was in the same position as Isaiah with his journey to heaven: it may be an effect of time dilatation.

As mentioned earlier, this factor of a difference of time in space is actually borne out by modern science. If a vehicle approaches relativistic velocities (near the speed of light), time appears to slow. It is not that time itself is actually changed, but that the distance traveled shortens, hence the "time" shrinkage. This means, in effect, that a vehicle traveling at nearly the speed of light to, say, the Centaurus system takes five years to get there and five years to return. Between the disappearance of the ship and its return again ten years will have elapsed, *to those awaiting it on earth.* To the crew, however, only several weeks will appear to have passed. It has been said that a circumnavigation of the galaxy at a minute fraction under the speed of light would be accomplished in half a million years' *elapsed time*, but could be completed in well under the lifetime of the crew.

We have already mentioned that a voyage to the Pleiades and back would take just under a thousand years, which connects reasonably well both with ancient legends about this star group being the home of the gods and with the thousand-year intervals between visits of the gods. This means that the God of Israel could make many voyages at thousand-year intervals, covering a span of many thousands of years, with the astronaut aging hardly at all. Such a traveler may well have seemed to the ancients not only immortal, but a "God"!

It certainly seems strange how ancient legends of space voyages and modern relativity theory converge; this must surely be more than a coincidence.

The second mythological factor giving credibility to a space civilization is the channels of communication with the gods.

The Phoenician *Sanchuniathon* (1193 B.C.) and *Philo Byblos* (A.D. 15) mention animated stones. The Christian historian Eusebius (c. A.D. 260–340) carried on his chest such a stone, which answered questions with a low whistling.

The Bible mentions "teraphim" or images which answered questions (Ezek. 21:21 and Gen. 31:34). Maimonides (1135–1204) in *Les Regeles des Mouers* states: "The worshipers of the teraphim claimed that, as the light of the stars filled the carved statue, it was put on a rapport with the intelligences of those distant stars and planets who used the statue as an instrument. It is in this manner that the teraphim taught people many useful arts and sciences."

Seldenus in *Die Diis Syriis* mentions teraphim consecrated to a special star or planet and says they were known to the Egyptians.

What were these speaking stones—stone or crystal devices that could pass messages from the stars? Do they not remind us of transistors, widely used in radio and TV today, which are crystals? Were they channels of communication akin to radio for communication with other worlds or with vehicles orbiting invisibly around our planet?

Again, we have a myth which is not quite so mythical when we relate it to modern techniques.

It is becoming less and less a fantasy that in a distant past, astronomical knowledge was at a high level, based on sophisticated observational techniques and on the possibility that there was space travel and contact with other worlds and intelligent life elsewhere in space. Perhaps one day our astronauts will see other, older footprints in the dust of the moon and know that men in a forgotten past also have trodden the road to the stars.

8

The Antiquity of Science

WE HAVE OBSERVED that there existed in the past a great deal of knowledge about physics and astronomy which could not have been gained without instrumentation and the resources of a technological civilization. We have suggested that repositories of the knowledge possessed by a superior culture were created when the civilization was threatened with extinction. Apart from the pyramids, which legends show were built against a catastrophe to hold the wisdom the human race had gained, there may have been many other such secret storehouses scattered around the world. Perhaps underneath the great artificial hill of Silbury in England there were such chambers, and the recently discovered underground chambers in South America and Turkey may also have held archives of scientific knowledge.

It also appears that some fragments of knowledge from a high culture survived in oral traditions, as witness the legend of Sirius's "dark companion" from the Dogons of the Sudan. By a quirk of fate it appears that this fragment was remembered by someone and the story handed down through the ages, while almost everything else had been forgotten. We have also suggested that the scientific knowledge of the ancient Greeks was gained from the great libraries in Egypt, and that this knowledge stemmed originally from what was saved from within the pyramids. There were a great many libraries in the ancient world, and little except scattered fragments of this literature has survived to the present day.

Apart from the libraries of Alexandria and the Temple of Ptah, there have been many "burnings of the books" in antiquity. The library of Pergamus in Asia Minor, for example, contained 200,000 books, all of which were destroyed.

When the Romans razed Carthage in the Punic Wars in 146 B.C. they also burned to ashes a library said to contain half a million volumes. The Romans also destroyed, under the leadership of Julius Caesar, the Druidic library at Autun in France, containing thousands of scrolls on philosophy, medicine, astronomy, and mathematics.

In China the Emperor Tsin Shi Hwang-ti ordered all the ancient books destroyed in 213 B.C.

Leo Isaurus burned 300,000 books in Byzantium in the eighth century A.D. The Catholic Church's Inquisition in the Middle Ages destroyed unknown numbers of heretical works of literature, which, because of their pagan origins, were anti-Christian and therefore "works of the devil."

It is probably true that medieval Christianity was responsible for the loss of irreplaceable knowledge and for holding the human mind in the bondage of a mental Dark Age for centuries. Not only did the priests eliminate great areas of forbidden knowledge in Europe, but in their voyages to the New World they destroyed the literature of ancient cultures far superior in many ways to the Europeans of the time. In the process two great civilizations, the Aztec and the Inca, were wiped out.

In Yucatán a young Spanish Catholic monk, Diego de Landa, discovered a large library of Maya manuscripts. "We burned them all because they contained nothing except superstition and the machinations of the devil," he wrote. To this day the surviving inscriptions of Maya glyphs have never been deciphered, except for some mathematical symbols, and those without any degree of absolute certainty. When we consider the extent of the fragmentary knowledge contained in one of the very few surviving works, the *Popol Vuh*, the bible of the Central American peoples, it is possible that the loss of knowledge was almost as great as that caused by the destruction of books in the Mediterrarean world.

Not all the blame, however, can be laid to the acts of the Asians and Europeans. When the Spaniards entered Peru they discovered a civilization highly advanced, well organized, and with a strong central government more efficient than any possessed by the Europeans of the time. They found well-maintained public works; excellent roads and bridges; systems of irrigation, water supply, and public buildings; regulated food production and storage, and supplies of all kinds carefully controlled in a balanced and efficient economy.

A huge, well-knit empire functioned perfectly under the Royal Inca, yet without any written records except those maintained by the system of quipu, or knotted strings, and without the use of a monetary system. The Inca had a superstitious fear of writing, and the explanation given to the Spaniards was that at a time of a great epidemic the oracle said that writing had to be done away with under the penalty of death. It appears that writing had been known to the earliest phases of Inca rule and was eliminated. This seems to be a parallel with the Chinese destruction of books in 213 B.C., when all the old knowledge was eliminated.

Was it really writing the Incas were afraid of, or of what the writings may have contained? If they were aware of the nature of the conflict that had destroyed a civilization and almost mankind itself, then perhaps they considered that the written knowledge posed a menace to future ages and should be destroyed.

Perhaps the long-vanished writings of the Incas held the same sort of knowledge as many Sanskrit texts which refer to nuclear weapons. This knowledge was possibly considered far too dangerous to survive, so who can blame the Incas? Even in our day, many of the scientists concerned with the making of the first nuclear bomb would have preferred this knowledge to have remained only in the realm of theory. At best, they hoped the device would not work in practice.

Was this Inca legend of a pestilence connected with atomic weapons and their effects? There are fused areas in South America, and also in North America, from where the ancestors of the Incas may have fled, which remind us strongly of the vitrified areas created by modern nuclear tests.

If it is thought that the Incas never possessed or knew of a written language, then we would be unable to explain the existence of a written language which has been discovered in South America. As we have seen, thousands of leaf-thin gold plates were discovered in deep underground caverns in Ecuador, and the language appears to have been neither glyphic nor pictorial, but *alphabetical.* Furthermore, this alphabetical writing bears a resemblance to ancient Cretan and Sanskrit scripts. Was this the type of writing which was eliminated from human memory by the Incas?

In this connection, it is of interest to observe that the Maya have a legend about their "Golden Book," supposedly a history of the Maya written on gold plates which were hidden at the time of the Conquest to prevent them falling into the hands of the Spaniards. The discovery of such plates in Ecuador gives credibility to the belief that the Maya were dealing in realities when they spoke of the "Golden Book," which has never been found. The fact that similar things have been found means that the Maya book could exist, and there are Maya legends about secret underground places that were built long ago. Perhaps one day further discoveries of this nature will come to light, but unless we find the key to decipher this unknown language, we may still learn nothing.

It seems that there was both a deliberate and a subconscious urge, throughout history, to eliminate ancient recorded knowledge. That which was not destroyed, either accidentally or by design in wars, was deliberately suppressed because it was *dangerous.* Possibly some of these attempts were well-meaning; it has not mattered in the final analysis. We have found our own road to atomic hell.

Unfortunately, however, the destruction of knowledge was not selective: the good was destroyed along with the bad. Most areas of knowledge—whether they be nuclear physics, the secret of flight, facts about explosives, or biology and medicine—have applications which are both good and bad. Perhaps those ancients considered that with such dangerous information the bad outweighed the advantages of the good.

The fact that ancient wisdom existed in many fields shows that the information that existed about, for example, astronomy, was not an isolated freak, but the inheritance of an advanced and balanced culture, close to our present-day civilization.

There exists a great storehouse of ancient science in the field of biology. Many of the cures claimed by the people known as witches, as well as medical remedies handed on by ancient rural practices, seem to stem from a very distant tradition. It is now known that the old folk cure for sepsis—the one using the mold of a potato or fruit—is actually a primitive version of penicillin. It seems that even in very early times the curative properties of molds was known, because in ancient Egypt a papyrus of the 11th Dynasty tells of a fungus growth on still water which possessed curative properties for wounds and open sores. (Ref. H. M. Botcher, *Miracle Drugs*, Heinemann, London, 1963.) Warmed soil and soya-bean curd, both of which have antibiotic properties, were used in ancient China and Greece.

Skulls unearthed from early Egypt have been found to contain artificial teeth, and dental crowns and fillings have been observed in skulls in Campeche, Mexico, which date from the Maya civilization.

Although most modern medical and surgical practices are based on knowledge which has come down to us from ancient Egypt, Babylon, and Greece, the medical and surgical abilities of the Peruvian coastal cultures was as great, if not greater, than those possessed by the Mediterranean peoples. Possibly the reason this fact has largely been overlooked or ignored is that these civilizations flourished in isolation from the rest of the world until the sixteenth century, and also the fact that no written treatise on their medical skills has been passed on to us. In pre-Columbian Peru we have no medical heroes such as Hippocrates to inspire Western medical techniques.

However, the decorations on a great many pottery vessels from the coastal Chimu culture have included remarkable portrayals of every aspect of Chimu life, including many on medical skills. It appears that these ancient people performed delicate operations: they amputated limbs, and fitted artificial legs, arms, and even hands. When one considers the excellence of wood carving from the Peruvian coastal region, it is not too surprising that they were able to make accurate models of human limbs. What is surprising is that they were able to carry out major amputations with such good results that the patients lived to use artificial limbs. Even in nineteenth-century Europe major amputation often resulted in death from shock, due to the crude methods and lack of anesthetics, or from septic conditions resulting from the ignorance of hygiene.

Some form of anesthetic must have been known to the Chimu surgeons, as pottery vessels show procedures being performed, including abdominal operations, which could scarcely have been carried out without a pain killer of some kind. In China, acupuncture has been used as anesthesia for an unknown period of time, and it is apparently effective even for major operations, both on human beings and on animals. Is it possible that this may have also been used by the South American peoples?

It has been suggested in recent years that contact between ancient China and the American pre-Columbian civilizations was not unknown. Some of the Mayan art forms and the traditions of many American Indian tribes, particularly of those from New Mexico, show an affinity with Chinese art and traditions. It has also been suggested that some of the Mexican glyphs found at Monte Alban and Teotehuacan have a relation to archaic Chinese. Contacts between these peoples in art and tradition would also mean an exchange of trade and knowledge. It is thus possible that acupuncture as a medical aid was not limited to ancient China.

Old Sanskrit texts reveal that ancient India had an extensive knowledge of medicine and understood the circulatory and nervous systems. Their doctors performed Caesarian operations, trepanned for brain surgery and used anesthetics. Sushruta in the fifth century B.C. listed the diagnosis of 1,120 diseases and described 121 surgical instruments.

A Brahmin book, the *Sactya Grantham,* described vaccination in 1500 B.C., centuries before the credit was given to Edward Jenner (1749–1823), as follows:

"Take on the tip of a knife the contents of the inflammation, inject it into the arm of a man, mixing it with his blood. A fever will follow, but the malady will pass very easily and will create no complications."

A magic mirror that could illuminate the bones of the body was said to be possessed by the Chinese Emperor Tsin-Shi (259–210 B.C.) and was kept in the palace of Hien-Yang in 206 B.C. The instrument was used for medical purposes. X rays in antiquity? Here one is reminded of a drawing found in Mexico, which shows a human figure with a rectangle over the chest area and within the rectangle there is a stylized drawing of what appears to be the spinal column and ribs. The significance of this has escaped the anthropologists and archaeologists, but the only logical explanation is that it is a drawing of an X-ray device in operation.

Here again, has this a connection with Chinese visits to the Americas, or does it once again demonstrate sophisticated medical techniques in ancient America? We know already the advanced state of medicine in the Americas, and for the surgical techniques we have observed among the Chimu some sort of X-ray device as an aid to diagnosis is virtually a necessity. But what of the belief that the Chimu, as well as other American pre-Columbian cultures, had no knowledge of electricity? Electricity would be essential if some sort of fluoroscopes were used. In this connection it is interesting to observe that the Chimu plated objects with gold in a way which today can only be done by electrolysis. We remember that ancient batteries were discovered in the Middle East, and it is not impossible that a form of electricity was also known in the New World.

The Aborigines of Australia have practiced blood transfusions for many thousands of years and their women employ oral contraceptives made from the resin of a particular plant. Apparently, unlike our artificially developed oral contraceptives, these have no harmful side effects, and they are chiefly used in periods of drought or food shortages, so that children are not born who cannot be fed. It is a sensible method of population control which many countries should copy.

The Aborigines pose something of a riddle to anthropologists, because there is some evidence that they are the remnant of a highly advanced people forced to live exceedingly primitive lives in a hostile environment. Possibly the above-mentioned medical skills are two things that have been remembered and passed down through tribal lore from a remote period, while almost all other early knowledge has been forgotten.

The Aborigines also possess a most useful weapon and hunting tool in the boomerang, which has been described as a very efficient aerodynamic shape that could scarcely have been invented by pure primitives. Here again, this may be an echo from an earlier epoch, when a weapon had to be invented, drawing upon scientific skills, for use by people deprived of all but the most basic means of survival.

There is no measure of agreement among anthropologists regarding the origin of the Aborigine, although it is generally thought that he was not indigenous to Australia. Estimates as to the time of his arrival vary from 20,000 years to 10,000 years ago. His racial stock is uncertain, but he has been described as an archaic Caucasoid, due to the heavy facial and body hair that is absent in both Mongol and Negroids. His dark coloring is considered a result of adaptation to hot desert conditions. The Aborigine has a level of intelligence not less than other peoples on earth; such a level of intelligence is required for him to adapt successfully to the hostile environment in which he lived.

Another aspect of ancient biological techniques concerns plant and animal husbandry. We have never with any certainty been able to identify the wild ancestors of corn or wheat. Indian corn, or maize, is known to have existed in a cultivated form in South America for more than four thousand years. Although unknown in the Old World at one time, maize now forms part of the staple diet of many peoples far removed from the area of its original cultivation.

Wheat is generally supposed to have been developed from a wild ancestor called emma, and it was first cultivated in the Middle East as long as 8,000 years ago. Perhaps this is not really so; what we call the wild ancestral form could be a degenerate species of a once-cultivated form which had grown wild and untended for thousands of years.

In South America the Incas were great agriculturalists and had developed many food plants. Some of the varieties they grew, which died out due to the destruction of the Inca empire, we today have been unable to duplicate. They grew maize, gourds and squashes, a great variety of beans, and many types of potatoes, one of which, a frost-resistant variety grown high in the Andes, has now vanished along with the secret of its cultivation.

The Maya today make soap from a tree called, appropriately, the soapberry tree, and they also produce honey from *stingless* bees. There is an old Maya legend that Quetzalcoatl, the culture god of Central America, developed cotton plants that grew different-colored fibers. Before we dismiss this as impossible, we have to remember the different-colored flowers that we have produced by selective breeding—perhaps something akin to this was done in the past.

The popular flower dahlia was brought to Europe by the Conquistadores from Mexico, where it was grown by the Aztecs in great profusion and variety; and no wild ancestor of the tuber has ever been located. If we cannot trace early versions of many of our most useful and ornamental plants, when could they ever be said to have been wild? And if they were never wild, where did they come from in the first place?

Were they imported, along with the colonists, from another world where they had been developed? Or were some of our domestic plants specially developed by biological engineering for use in a new planetary environment?

We have developed many useful plants from wild ancestors by selective breeding and careful cultivation. This brings us to another puzzle. We have seen that the master blueprint for all forms of life is contained within the DNA structure, which determines all the possible characteristics of each particular form. Therefore, all the variations inherent in any particular species are already coded in the DNA, and selective breeding (which is how plants are improved) is done by selecting certain characteristics which are *already there.*

We cannot create a new plant for which the potential does not exist in the DNA, which is probably why we have been unable to produce a true black rose or black daffodil. If the potential for this characteristic does not exist within the DNA molecule of the particular plant, then we shall forever be unable to produce it. Therefore, no matter how many variations we produce from a "wild" ancestor to a cultivated state, we are only selectively using capabilities which already exist but are not utilized, either because they are not necessary for survival in a wild state or because they detract from useful survival characteristics.

Looking at it from this point of view, are cultivated plants descendants not of wild plants but of plants once domesticated—which reverted to the wild state, their specialized characteristics becoming dormant over the ages to better enable them to survive in a hostile environment? This idea again connects with the ancient concept of a "Garden Paradise" which later reverted to a wilderness. The development of plant cultivation among both the Inca and the pre-Inca peoples presupposes a long period of development, *or* the inheritance of a great fund of knowledge in plant husbandry and genetics.

If such knowledge was not inherited from a scientific past, how were special products known to the ancients, or to present-day primitive peoples?

Not only do the Australian Aborigines use a plantderived oral contraceptive, but the forest natives of New Guinea use similar herbs to the same end. Amazon Indians pulp the leaves of a certain plant to use as a very efficient antibiotic against jungle ulcers, which even penicillin will not wholly cure. Did they evolve these methods by trial and error over a long period of time, and if so, what idea led them to experiment in this manner? It would seem that by chance alone they would never have stumbled across the properties of many of these plants. Were their scientific ancestors aware of these properties by the use of analytical methods before they were thrust into a savage and primitive environment? Was the information inherited by oral tradition, in the same way that the Dogon of the Sudan had certain odd knowledge of the stars?

It has often been said that the great divergence between cultivated plants in the Old World and in the New points to a completely separate development of civilization in both hemispheres. It is held that if there had been transoceanic contacts between the two civilizations many more plants would have been spread to both sides of the Atlantic. Maize, for example, would have been grown in Asia and Africa thousands of years ago as well as in America. The people would have carried seeds and plants with them by way of trade. This is, of course, supposing that the world was in a fairly primitive state.

However, if we assume that there had existed a world civilization comparable to our own, then this factor of separate growth areas may not have been as significant. For example, today the staple drink of England is tea, vast quantities of which are used each year—but the tea plant itself cannot grow in the temperate climate of England, except under artificial and experimental conditions which are not commercially viable.

A future plant biologist, therefore, thousands of years after our civilization had ceased to exist, probably would never suspect that the English drank tea, because the plant for its production *did not grow in that region.*

In the same way, an advanced civilization of the past may have selected certain areas where certain plants were best adapted and intensively cultivated them for distribution *around the world.* Being extremely perishable, it is almost certain these plants would leave no traces in the country of their importation, especially after a long period of time.

The great differences in diseases among pre-Columbian civilizations and those of the Old World have been cited as another argument for separate development. Many European diseases were completely unknown to both the American Indians and the Polynesians until contact with Europeans and, because of their lack of resistance, whole native populations were decimated. However, if all diseases had developed naturally, and microorganisms were part of the planetary ecology, there should be no reason why one continental mass should suffer some diseases and the other escape.

But if disease was loosed artificially—we have previously mentioned the possibility of bacteriological warfare—then it is possible that either this form of warfare was geographically limited, or some populations were able to prepare forms of defense against it. Admittedly, answers of this kind to the separate-development hypothesis are tentative, but they are possibilities, particularly if we assume that a highly developed civilization existed in the remote past.

Turning now to the fields of more mechanical inventiveness, we find evidence of many things commonplace in our twentieth century which appear to have existed in the remote past. Descriptions by ancient writers and certain aspects of mythology lead us to the supposition that some form of artificial lighting, perhaps electric, existed. Records indicate the existence of extremely accurate mapmaking (which requires advanced surveying techniques and instrumentation), robots, microscopes, and telescopes. We have already mentioned great astronomical knowledge in the past, as well as the possibilities of radio, radar, and aircraft.

Electricity is supposed to be a recent invention, but we have already seen that for the ancients to have possessed X-ray equipment there must have existed a form of electricity, and the possession of this form of power would explain the gold plating which cannot be explained by any other means.

The German archaeologist Wilhelm Konig found near Baghdad in 1938–39 a number of earthenware jars with necks covered with asphalt and iron rods encased within copper cylinders. He thought they were ancient electric batteries.

After World War II Willard Gray of General Electric made

duplicates of the ancient batteries, using copper sulphate instead of the unknown electrolyte which had dissipated, and the battery worked. This seemed to point to the fact that the ancient Babylonians knew electricity. Electroplated articles have been unearthed in the same general area, which supports this view.

Professor Dennis Saurat has found what he considers to be the remains of electrical devices in ancient Egypt. The rock tombs of Abu Simbel had always posed a great puzzle: the interior chambers, cut from the solid rock, were painted in great detail —yet no trace of fire blackening has been found, as would be the case if ordinary torchlight had been used. Indeed, no trace of blackening from torches has been found in any Egyptian building where artificial illumination would be needed. Did, then, the ancient Egyptians use electric torches, perhaps carefully preserved from an earlier age?

As further evidence, ever-burning lamps are frequently mentioned in historical documents. Numa Pompilius, king of Rome, had a perpetual lamp shining in the dome of a temple. Plutarch mentioned a lamp at the entrance to a temple of Jupiter-Ammon, whose priests claimed it had burned for centuries.

Pausanius described a golden lamp that could burn for a year in the Temple of Minerva. St. Augustine (A.D. 354–430) told of a temple dedicated to Isis where there was a lamp that neither wind nor water could extinguish. This description is remarkably like that of an electric lamp which cannot be affected by wind or water, unlike an ordinary flame torch. An inscription indicated that a lamp at Antioch during the reign of Justinian (sixth century A.D.) had burned for five centuries. In the Middle Ages a perpetual lamp was found in England that had been functioning for several centuries.

There is an old Sanskrit text called the *Agastya Samhita* which gave instructions for making batteries:

"Place a well-cleaned copper plate in an earthenware vessel. Cover it first with copper sulphate and then by moist sawdust. After that put a mercury-amalgamated zinc sheet on top of the sawdust to avoid polarization. The contact will produce an energy known by the twin name of Mitra-Varuna. Water will be split into Pranavayu and Udanavayu. A chain of one hundred jars is said to give a very active and effective force."

Mitra-Varuna is now cathode-anode. Pranavayu and Udan-avayu are oxygen and hydrogen respectively.

Electricity in Europe, the Middle, and Far East in an ancient era. Fables? The same stories are reflected on the other side of the world in the Americas.

The Maya have stories of cities that were lit by day and night. Many ancient Maya buildings have no windows, and yet in those also, as in the ancient Egyptian structures, there is no trace of fire blackening inside these painted chambers. Even the Amazon Indians have legends of the cities of ancient times that were lit by "stars"—electric lights? These stories were told the Spaniards hundreds of years ago, so this is not a modern legend which has arisen among jungle tribes in recent times.

Gold objects plated by methods which today can only be done by electrolysis have been found, notably in the area of the old Chimu capital, Chan-Chan. Also in the Americas, there have been discovered objects made of platinum (which the Spaniards discarded as useless). This metal requires temperatures of 9000° centigrade for smelting, and temperatures of this range cannot be produced by ordinary fires.

In Greek legend Hephaestos, the blacksmith of the Olympian gods, had two golden maidens for servants. "They are made of metal," he explained, "but they do my bidding, and have thoughts in their heads." Robots? It will be remembered that it was Hephaestos who built Talos, the metal giant who guarded Crete and who threatened anyone who landed. Was this also a robot, programed to carry out certain tasks?

When Jason attacked the heel of Talos, was he perhaps switching its power supply off or disconnecting it in some manner? If these robots had been built by one of the Greek gods, then they must be presumed to stem from a remoter era than the Greek civilization. Perhaps they were powered by solar cells and functioned for centuries. Once they broke down, in a less scientific era, they would be broken up and melted for spearheads.

In China also there are reports of mechanical men in antiquity. The Emperor Ta-chouan possessed one, and the jealous emperor had it destroyed because the empress coveted it.

Consider the curious phenomena in ancient times of the oracles, objects that could speak wisdom and answer questions. There was the Oracle of Delphi in Greece, and there were temples in Egypt where the statues could speak. No doubt some of it was trickery, the work of hidden priests, but is it also not possible that some of these things may refer to ancient computers which were still functioning centuries after their makers became dust?

Garsilaso de la Vega was told that the Incas had a statue in the valley of Rimac "which spoke and gave answers to questions like the Delphic Apollo." Could this perhaps have been a form of computer?

If an ancient, highly advanced culture had collapsed, among the measures taken to preserve a degree of knowledge may have been the installation of computers programed with information useful to the survivors struggling back to civilization. It is not unlikely that we ourselves, should our civilization be menaced with almost total destruction in, say, a nuclear war, would prepare and program computers with information for those survivors who would need it.

Sponge divers near the island of Antikythera in 1900 found a metallic object which, when it was cleaned, was found to consist of a system of complex cogs and gears. Dr. Derek J. de Solla Price, a scientist working at the Institute for Advanced Studies at Princeton who examined the object, said:

"It appears that this was, indeed, a computing machine that could work out and exhibit the motions of the sun and moon and probably also the planets."

This device, believed to have been made about 65 B.C., came as a surprise to modern scientists—nothing like it had been discovered before among Grecian artifacts, yet it must have had a long history of development. "Finding a thing like this is like finding a jet plane in the tomb of King Tutankhamen," said Dr. Price at a meeting in Washington in 1959.

This calculating machine was found in the sea among amphorae, which may lead us to suspect it was being carried in a ship which was wrecked. We can ask ourselves: If this was being carried in a ship, was it being taken somewhere *from* Greece, or was it being *brought to Greece* from somewhere?

It may be pertinent to note that the instrument was designed primarily to measure the movements of the sun and the moon—we are not certain that it calculated the movement of the planets. We suggested in *Colony: Earth* that Stonehenge was a computer built for a specific purpose: to study the movements of the *earth, moon, and sun only*. We held that Stonehenge was not built by the Greeks (as has often been thought) but that the Greeks went there to study these motions and learn from the Stonehenge mathematicians. Was the computer, then, either built in Britain or based on data obtained from the Stonehenge computer?

This would mean that Egypt was not the only place where the Greek students went to gather their knowledge, but that much of their mathematical expertise was in fact learned from the wise men who lived in the far north land of the Hyperboreans, where there was a round temple of astronomy and mathematics. The fact that the Greeks frequently alluded to this northern land points to Britain and Stonehenge.

9

Horizons of Yesterday

IN LATE medieval times in Europe, the world was a curiously shrunken place. The geography of Europe and the Mediterranean was known, although not to the mass of the people, and there was a vague awareness of a country called India and another called Cathay, and even of some islands off its coast. But no maps were being drawn which were even approximately accurate. Europeans were unaware, at that time, of the islands of the Pacific or the continent of Australia, the island of New Zealand or the southern part of Africa. They knew a little of China from the tales brought back by Marco Polo; of the Americas and the advanced civilizations and great stone cities that existed there they were totally ignorant until the opening of the New World in the late fifteenth and sixteenth centuries by the Spaniards.

Most Europeans of the period thought that the earth was flat and that to sail too far across the Atlantic Ocean was to face certain death by falling over the edge. Some thought that far out at sea was a belt of fire girdling the earth. So until Columbus and Vespucci sailed in their little ships, the explorations of the Europeans were severely limited.

Further back, in late Roman times, there was even less knowledge. Civilized men of the time basically viewed the whole world as encompassed by the Mediterranean and the lands surrounding it, although India's western coast was known to the Romans. Britain was familiar, of course, and there was a suspicion that there were colder lands farther north where the sea rovers lived, but none went there. There were even more vague stories of islands that lay over the great ocean, but there were neither the ships or the men to venture so far away.

Yet, surprisingly, the further back we go in time, the *wider* these shrunken horizons become, which surely is the opposite to what it should be.

The ancient Egyptians circumnavigated Africa, although this was forgotten by the time the Romans had conquered Egypt.

The ancient Greeks apparently either knew or suspected that other great lands existed apart from the known Mediterranean world. Once again, can we not think that this was knowledge gained from their studies in the archives of Egypt?

Philostratus of Athens (A.D. 175–249) said: "If the land be considered in relation to the entire mass of water, we can show that the earth is the lesser of the two." This points to a knowledge of the comparative world masses of land and sea—could it possibly have been gained by a study of ancient maps?

Plato said in the *Phaedo* that the peoples of the Mediterranean occupied only a small portion of the earth.

Pytheas of Marseilles (330 B.C.) voyaged to the Arctic Circle and described the midnight sun.

Seneca may have heard of the Americas, because a verse in *Medea* says:

> There shall come a time
> When the bands of Ocean
> Shall be loosened
> And the vast Earth shall be laid open.
> Another Tiphys shall disclose new worlds
> And lands shall be seen beyond Thule.

Thule was the ancient name for Iceland, and new lands beyond must obviously refer to North America.

The *Vishnu Purana*, a Sanskrit text, describes a Pushkar (continent) with two Varshas (lands) at the foot of Meru (North Pole) and says this continent faces Kshira (an ocean of milk) and the two lands are shaped like a bow. What sounds like a mythical fairy tale is in reality quite a reasonable description of the Americas. The two lands (North and South America) face the Arctic Ocean (the Ocean of Milk—presumably because of the ice); they lie under the North Pole, and their shape could be described as a bow.

Like so much else in ancient texts from India, this again seems to be knowledge passed down from a very remote period, for the Hindus of two thousand years ago did not have seagoing vessels capable of making the voyage to the Americas. No doubt the Brahmin priests of that time did not themselves understand the meaning of this description, but as with the mathematical texts they merely copied faithfully the sacred writings.

The Bon sect of Tibet possesses an ancient book which contains a chart marked off into squares and rectangles with the names of unknown countries. The Soviet philologist Bronislav Kuznetsov came to the conclusion that the chart was a map, and identified the Persian city of Pasargady (fourth to seventh centuries B.C.), Alexandria, Jerusalem, the countries of Bactria, Babylonia, North Persia, and the Caspian Sea. So the remote Tibetans, locked within their high mountain fastness, had a great deal of geographical knowledge many centuries ago.

The most extraordinary discoveries made in recent years, however, have been the maps of Admiral Piri Reis, two of which, dated 1513 and 1528, have been verified as genuine by both the American Geographical Society and the Royal Geographical Society of Great Britain.

The map dated 1513 shows Brittany, Spain, West Africa, the Atlantic, part of North America, and a complete outline of the eastern half of South America. At the bottom of the map is shown the coastline of Antarctica, eastward to a point under Africa.

The second map, dated 1528, shows Greenland, Labrador, part of Canada and the east coast of North America down to Florida.

It is suspected that these two maps are but part of a world

mapping series, perhaps showing the Indian Ocean and the entire Pacific region, although these have never been discovered.

There is another map, belonging to Orontus Finaeus, dated 1531, which also shows the outline of the Antarctic continent. This even includes rivers and mountain ranges.

In Iceland was discovered the map of Zeno, which shows Greenland as three islands under the ice cap. Rivers and mountains are clearly marked. This map is dated even earlier than the maps of the Americas and Antarctica—1380.

There are many peculiarities about these maps, which in themselves indicate a totally different knowledge of the world in antiquity than had been suspected, and also suggest a very different world from the one we live in at the present time.

First, the Piri Reis maps. The Turkish admiral Piri Reis compiled an atlas called the *Book of the Seas* containing 210 well-drawn maps. The origin of the two Antarctic maps is amazing. According to notes made by Piri Reis, his uncle, Captain Kemel, captured a Spanish sailor during the course of a naval battle. This sailor had in his possession some rare maps. He said that these were the ones used by Columbus in his discovery of the New World, and that this sailor had in fact been on three of Columbus's expeditions. The sailor said, under questioning: "A certain book from the time of Alexander the Great was translated in Europe, and after reading it Christopher Columbus went and discovered the Antilles with the vessels he obtained from the Spanish government."

It appears, therefore, that Columbus did not sail blindly across the Atlantic Ocean in his quest for the New World, but that he possessed extremely accurate maps, so that he knew exactly where he was going. This is further borne out by the fact that Columbus described the shape of the world as slightly pear-shaped. As we have only recently discovered this fact ourselves from our voyages in space, he must have obtained this information from a very ancient source.

The problem is this: These maps from the time of Alexander the Great could not have been originals, because the knowledge of the Greeks then was insufficient for them to have drawn up such maps.

They neither traveled such distances nor had the necessary surveying equipment. We can only conclude that the maps were copies of charts even more remotely ancient. Once again, does it not seem possible that this is further knowledge which some unknown Greek student had gleaned from the libraries of Alexandria, which were said to contain all the knowledge of the human race from a remote time?

Examination of the maps has shown them to be incredibly accurate. The American cartographer Arlington H. Mallery, with the cooperation of the U.S. Navy Hydrographic Office, determined that the distances between Europe, Africa, and the Americas on the Piri Reis maps were exact. This despite the fact that until the eighteenth century no longitudes had been determined with any accuracy.

The chart of South America shows the rivers Orinoco, Amazon, Parana, Uruguay, and Rio de la Plata with extreme accuracy, and none of the explorers of the southern Atlantic at this time—Vespucci, Magellan, or Columbus—had charted the rivers of South America beyond the coastal deltas. Therefore we can see that the continents of North and South America were mapped with extreme accuracy, not only before the time of the fifteenth century explorers, *but before the time of Alexander the Great.*

It is when we look at the portions dealing with Antarctica that the greatest mysteries lie. Both the Piri Reis maps and that of Orontus Finaeus accurately chart the outline of the continent and show rivers and mountain ranges. For one mystery, the first explorations of Antarctica did not begin until the nineteenth century, and no really accurate mapping was completed until the International Geophysical Year 1951. When the survey was completed the old map was compared with the seismic soundings through the ice cover taken in 1951. A discrepancy was discovered and a second seismic probe undertaken. It was then discovered that an error had been made—but it was not the ancient Piri Reis map that was wrong. The old map was more accurate than the modern ones.

The other peculiarity about the Antarctic maps is that they show mountain ranges and river courses whose existence was not suspected, and that were not charted until the 1951 I.G.Y. The fact that the ancient maps show the continent that is now hidden by ice leads to the conclusion, unwelcome though it may be to many scientists, that the maps were drawn at a time when there was *no ice at the South Pole.*

The map of Zeno, showing Greenland, was likewise checked by the French polar expedition of Paul-Emile Victor in 1947–49. Again, the ancient map was found to be extraordinarily accurate and, like the maps of Antarctica, had been apparently drawn when the North Polar regions were ice-free.

This must surely be a hard pill for the scientists to swallow. No matter which viewpoint is taken, these maps totally contradict all traditional scientific thinking regarding the earth's climatic history. Most scientists have been of the opinion—and still are—that the two polar regions have been sealed under the ice for many hundreds of thousands of years. Now, on the one hand, they can accept the view that the maps were drawn up hundreds of thousands or even over a million years ago—before the human race appeared on the planet, according to the anthropologists.

On the other hand, if they admit that human beings intelligent enough, and with a high enough level of civilization to draw up such accurate maps, existed so long ago, then their hypothesis of an Ice Age and the polar ice caps being in existence at this time are false. And how do they explain that cores taken from the sediments of the Ross Sea in Antarctica indicate that vegetation of a sub-tropical nature was growing at the South Pole some six thousand years ago, and fossils of orange and magnolia trees in the North Polar regions also indicate a much warmer climate with *no glaciation* in this area only a few thousand years ago?

These maps alone should cause a revolution in considering the past, and they point to two factors in particular that have never been taken into account by our scientists. One is that our ideas regarding the earth's climate in the past of six to eight thousand years ago are in error, and that quite possibly the hypothesis of an Ice Age in the past is totally erroneous—as was pointed out in *Colony: Earth.*

The second factor is that the ancient Greeks at the time of Alexander the Great were incapable of drawing up such accurate charts, and certainly no one later than this was capable either. Therefore they must have been drawn up at an earlier date, which means that a level of culture existed then that has disappeared. This also is completely at variance with the hypotheses at present in vogue among cultural anthropologists regarding man's state of advancement during the past ten or twelve thousand years.

Just how could these amazingly accurate maps have been made? Arlington Mallery says: "We don't know how they could map it so accurately without an airplane."

We have already discussed mentions of aircraft at a prehistoric period about which little is known apart from legends, and it appears that these maps must stem from that period. From such accurate maps—even more accurate than modern ones—a lot can be inferred. The people who undertook this mapping must have been in possession of advanced surveying equipment. For us to map as accurately we need to resort to photography from aircraft; therefore we can assume that the ancient mapmakers did likewise.

Also, did the surveyors travel to Antarctica from bases elsewhere, which means long-distance aircraft or powerful oceangoing ships? If it is true that Antarctica was free of ice at this time it could also mean that they operated from bases within the continent itself, which could lead us to suspect that Antarctica was inhabited at that point in time. It is perhaps significant to note that Piri Reis said that the atlas contained charts of all *inhabited* lands, which must therefore include the present South Polar regions.

We suggest that many surprises await us in the Antarctic continent. It is more than likely that if scientists ever are able to excavate the land which now lies under thousands of feet of ice they may discover not only traces of human habitation from the legendary past, but the ruins of cities whose existence has never been suspected. Perhaps Antarctica was one of the main continents where the ancient civilization of marvels, described in the old legends, reached its full flowering. The evidence of the maps indicates a complex, organized, and highly advanced technology at least on a level with our own, for such knowledge would not develop in isolation, but only as part of an advanced civilization.

It is beginning to appear that at one time the whole planet was as accurately mapped as it is today, for we have in our possession ancient detailed charts of the entire western half of the globe. Although we have no maps of comparable accuracy of the Asian continents, it is surely not mere fantasy to suspect that they also must have existed, even though they have not been found. There is always the possibility that they may be discovered one day. On the other hand, the existing maps may have survived the ravages of time by a fluke of fate, and the others may have disappeared forever.

Even so, the existence of these maps is further evidence for something suggested earlier in this book: If the whole world had been accurately mapped, it points to the existence of a *worldwide civilization*. This would account for the widespread legend of a Golden Age, and explain why researchers seek in vain for an isolated location of the original "Atlantis" or mother culture.

It may well be that, if our civilization is destroyed in a nuclear disaster, sages from future times will in their turn seek in vain for their Atlantis, being ignorant of the fact that the twentieth-century technological culture had spread all over the planet.

It is not the fault of the Mediterranean peoples or those of Northern Europe that their horizons shrank so drastically during the past four thousand years or so. The civilizations of the Americas were likewise unaware of peoples on the other side of the great ocean barriers. They possessed legends of white teachers who came from far across the sea, but they did not appear to wonder from what distant lands they might have come, or make any attempt to undertake voyages to discover these distant lands.

Of course, the Peruvian peoples such as the Chimu and the Inca lived on the Pacific coast, and the Toltec and Aztec lived far inland in Mexico, but even that other advanced race of the Americas, the Maya, many of whose towns faced the Atlantic, never attempted ocean voyages. Of all the peoples of the Americas, it was the Maya who may have sailed in the Atlantic, but they never produced an American Columbus to sail eastward to discover a New World in Europe or Africa.

We can now postulate that during a period we term prehistoric there existed a civilization whose members were aware of the shape of the earth and the distribution of the masses of land and sea. No doubt they were aware of all the lands that existed and all the major foci of population, in the same way that the average educated man today has at least general knowledge of continental lands and distances and the distribution of the world's major cities. The fact that this knowledge was almost completely forgotten can be best explained by a catastrophe which wiped out civilization on a *planetary scale.*

It can be demonstrated how easily civilization can be completely disrupted by taking as an example the destruction of the Roman Empire. With the collapse of Roman control in the fifth century A.D., outlying provinces such as Britain quickly lapsed into a state of barbarism. Within several centuries the Roman presence was almost forgotten—so much so that by the time of the Norman conquest the remains of the Roman roads were regarded with awe by the inhabitants and were thought to have been built by magic. Such constructions were so totally beyond the capabilities of the primitive society which had taken the place of Roman civilization that they could not imagine these roads to have been the work of ordinary mortals.

The destruction caused by invading bands of Northmen was so extensive that today scarcely anything survives in Britain of the scores of Roman towns that were built during nearly four hundred years of Roman occupation. Had we not a comprehensive documentary history of the Roman presence it would be difficult to realize that Britain was so thoroughly Romanized, had we to judge by the relative sparseness of archaeological remains alone. The fact that Roman civilization was so quickly and thoroughly forgotten is more surprising in view of the fact that the continuity of culture was never completely disrupted. Centers of learning survived all the hazards of the Dark Ages both in Britain and Europe, and, of course, the Renaissance was merely the rediscovery of knowledge that had survived the collapse of the Roman Empire and was brought to light when Constantinople was finally liberated from the Turkish yoke.

The effects on an advanced civilization of a planetary disaster on a vast scale, with massive geological upheavals and climatic changes, would have been much greater than the disruption caused by the collapse of the Roman Empire, and the scattering of the survivors into small isolated pockets separated by great distances would have hastened the process.

We have perhaps a pointer signifying a great catastrophe in the distant past if we refer to statistics which have been made relating to world populations. It has been estimated that there were some 250 million people, two thousand years ago. In 4800 B.C. there were 20 million. For 5000 B.C. the figure is given as 10 million, and in 6000 B.C. it has been reckoned that there were only 5 million human beings inhabiting the entire planet. At a period some time before 4000 B.C. the population of Britain was reckoned in the thousands, or even as low as the hundreds, and these in small, scattered groups.

It was suggested in *Colony: Earth* that the megalithic complexes in Britain (such as the chambered tombs) were built at this period, prior to 4000 B.C., although the archaeologists have generally stated they were built in 1000 B.C. Recently these dates have been reevaluated, and a figure *earlier than 4000* B.C. has been suggested. This remarkable aboutface has come about by reevaluation of the radiocarbon dating methods. By checking against trees of known age—notably the bristlecone pine that grows in California, some of which are known to be over *six thousand years* old—it has been found that all previously calculated radiocarbon dates prior to 1000 B.C. are too young. The date for Stonehenge has also been pushed further back into the past, to 2000 B.C., but in the opinion of the author, Stonehenge is almost as old as the chambered tombs and is at least a thousand years older than even the new date assigned to it.

These new dates will also require a reevaluation of the construction methods of the megaliths of Western Europe, in view of the small populations involved. It seems obvious that we shall have to seriously reconsider the methods suggested for their construction: muscle power by thousands of laborers, who, we now believe, simply did not exist.

We have these estimates of small populations in the past—as low as a few million some eight thousand years ago—yet the legends from various parts of the world point to great populations before the Flood. Genesis says: "And mankind spread all over the face of the earth." Ancient Sanskrit texts speak of great cities with *60 million* inhabitants being destroyed in one night. The *Popol Vuh* of the Maya says that "the lands of the Ten Regions were shaken and torn asunder and the cities with their millions of inhabitants sank beneath the sea."

Therefore, at a point about 8000 years ago, we have stories of populations running into many millions, and later populations are drastically shrunken. It begins to appear that the idea of tens or even hundreds of millions of human beings being killed in an awesome disaster is not so far-fetched. The destruction of the majority of the human race would certainly have thrust those who survived back into a barbarism from which it would have taken centuries to recover. Even the collapse of the Roman Empire never incurred such a disastrous loss of life as this, and we are all aware of the great dislocatixn of civilization that stemmed from this event, and the centuries that it took for Europe to recover.

It seems that the ancient Egyptians were aware of the destruction of former cultures and the setbacks to progress this caused. The Egyptian priests told Solon that the destruction of a civilization by a great calamity would also destroy that culture's knowledge. They said that those who would survive would perhaps be rude shepherds guarding their flocks high on mountaintops, while the cities which contained both the knowledge and the scholars would be lost to the fury of the elements. "You have to begin all over again as children," a priest told Solon.

In the *Timaeus*, Plato recorded the words of an Egyptian priest: "There have been, and there will be again, many destructions of mankind."

It is perfectly logical to assume that if there had been a nuclear holocaust in the past, all the major cities with their huge populations would be the first targets for the weapons. It is in the cities where human knowledge is concentrated—here are the great libraries and universities, and the centers of wisdom gathered throughout the ages.

In the countryside the simpler people lived, the agriculturalists and herdsmen. It seems probable that, no matter how advanced a civilization is, there will always be groups of people who are not concerned with knowledge as such, but who have simpler interests and more mundane pursuits. There is a world of difference between the space scientist and the humble farm laborer. Their activities, their knowledge, their outlook and spheres of interest are poles apart, yet they both are part of the same culture. But if you destroy all the scientists and technicians and only leave the farm workers to survive, then a rural civilization will develop, and what primitive technology there is will be geared to agricultural activities.

Of course, there will always be some intellectuals who would survive, especially in the case of a nuclear or other great catastrophe where there would be advance warning. Steps would be taken to ensure the safety of some intellectuals, and others would survive by chance.

The fact that this may have been the case in a former disaster is borne out by two circumstances which have long puzzled scholars. One is the sudden appearance, after the Flood, of gods, or "culture bearers," who led the rude survivors back to civilization by teaching them the rudiments of building, mathematics, medicine, agriculture, etc. This could have been the result of a deliberate policy to ensure that some educated people survived the catastrophe so that civilization could be rebuilt.

The second puzzle is the often uneven way in which some of the ancient civilizations developed. Most of them had the major attributes of civilized living, with reasonably well-planned cities, with proper water supply and drainage, organized government, efficient agriculture. Some of them, however, excelled in certain narrow fields: the Maya were master mathematicians of the ancient world; the Incas had the most highly organized centralized government; the Chimu possibly the most able surgeons; the Greeks excelled at artistic sculpture and abstract thought. It rather appears that there may originally have been experts in certain fields who had survived among the isolated groups, and so a certain talent was highly developed in that community and not elsewhere.

10

Further Oddities

THERE ARE many things in this world of ours that defy all the traditional explanations which have been offered, but which must have an explanation—and if the old answers do not fit, then we must find new ones, no matter how discomfiting to the experts. A theory is valid only if it fits the fact; new facts are constantly being brought to light that should rightly refute existing theories, but there is a stubborn refusal to change even when new evidence ought to make us think anew.

One of the most extraordinary discoveries in recent years has been made in South America. Dr. Daniel Ruzo in 1952 discovered a number of megalithic sculptures at Marcahausi, some fifty miles northeast of Lima, Peru. Marcahausi is a plateau roughly twelve thousand feet above sea level. It is always cold, and hardly anything grows in this stony and hostile environment. What faced Ruzo in a desolate rocky amphitheater was, by accepted standards, unbelievable. There were enormous figures of people carved in the rock, and the faces displayed the main groups of mankind: Mongoloid faces, Caucasian, and Negro. There were carvings of lions, elephants, horses, cows, and camels. He saw a representation of the amphichelydia, an extinct ancestor of the turtle known only from fossils.

Most experts are of the opinion that the major influx of human beings into the Americas were Mongols who had filtered through from Asia across the Bering Strait some twelve thousand years ago. This would mean that humanity had been occupying the Americas for only the past ten or twelve thousand years and these people were mainly of Mongoloid descent. This concept has been challenged in recent years, as traces of human beings contemporary with the oldest finds in Europe have been found, dating back twenty or thirty thousand years. Furthermore, there is another basis for challenging the concept of Mongolian colonization alone, because the remains of mummified corpses many thousands of years old have been found in Peruvian graves with wavy auburn or blond hair, which identifies them as of Caucasoid descent.

The Olmec carvings of giant heads found in Central America have a distinctly Negroid cast, although the anthropologist would deny that the Negroes ever lived on the American continent until brought there as slaves in recent times.

These carvings, however, show that someone in the past knew of these three races, and it is possible that all the types of human beings lived at one time in the Americas.

It is known that the horse was extinct in the Americas for nine thousand years until reintroduced by the Spaniards in the sixteenth century, which seems to date these carvings to a very remote period. This is further borne out by an analysis of the rock from which these carvings were made. Geologists estimate that the period of time which would be required for the incisions to take on the grayish tint in the white diorite porphyry would be at least ten thousand years. It would appear, therefore, that these carvings were executed at least ten thousand years ago—8000 B.C.

There is another curiosity regarding the sculptures of Marcahausi, and that is that some figures can be seen at different angles of sunlight, while others cannot. The figures change, appear, or disappear with the angle of sunlight and shadows. The faces show different aspects at different times—in one light, one takes on the appearance of age, at other times the same face is that of a young man. Obviously, the sculptors had a great deal of knowledge in the field of perspective drawing and optics.

We may know the *when* of these strange figures, but we do not know the *who* and the *why*.

A guess can be hazarded as to the who, although we can give no name to the civilization which vanished so cataclysmically before the dawn of recorded history. Some aspects of Marcahausi art are not isolated on this plateau in South America. Strange optical effects have been noticed in incised carvings in many parts of the world. The monoliths and other megalithic structures from Ireland—New Grange springs to mind—and from England and Europe are often carved with spirals, concentric circles, and other patterns. Some of these can only be seen in certain lights, at certain times of the day. Sometimes photographs reveal something different from that which the eye observes.

Since these display the same optical effects as the figures at Marcahausi, it is therefore not beyond the realm of possibility that the same civilization which carved the one, carved the other. This is not to say that the same people traveled throughout the world executing these works, but that the *same technology was employed.* As the optical carving effects are widely distributed, this once again points to a world-spanning civilization at a remote period.

This concept had also occurred to Daniel Ruzo, as he says:

"Hundreds of discoveries and observations of this kind made in South America convinced me absolutely that these sculptures could not possibly be mere freaks of nature, but must be the purposive work of a people whose civilization is as yet unknown. I called it the 'Masma Civilization.'

"All the works of this people had points in common—anthropomorphic and zoomorphic represenations, executed within a restricted space; repetition of the main themes, a combination of different designs on the same piece of rock; the complete effect visible only on a particular date.

"Between 1953 and 1958 I sent a number of reports to the Academies of Lima, Mexico and Paris about this.

"I observed the same effects in England at Stonehenge and at

Avebury, where one of the finest Druidic temples in Europe is to be seen. Careful examination of the enormous blocks of stone led me to believe that they had at one time been sculptured.

"One can only conclude that certain artists, whose origin is wrapped in mystery but who were no doubt trained to a kind of four-dimensional form of art, had for thousands of years carried on their function as sculptors for the Masma Civilization."

Ruzo adds: "In conclusion, I am of the opinion that in these places we have witness of a vast civilization that spread over the whole earth in the days before the Flood, but which I was unable to analyze in greater detail. . . ."

The author does not agree on all points with Dr. Ruzo, particularly as the Stonehenge and Avebury monuments are far older than any Druidic institutions, and although they may have been used by them at a later date, were not built by the Druids. But he supports the view (in a slightly different way) that these sculptures represent one facet of a highly advanced civilization from an unknown epoch.

Why were carvings made in South America of animals which no longer live there, to baffle future ages? Was it that all these creatures lived in South America at a remote period, even though we have been unable to trace their fossilized remains in all cases? Were these drawings created to show future ages that things were once very different from the present?

Perhaps there was another reason. Mount Rushmore in the United States has had a series of gigantic carvings of the heads of American presidents cut into the cliff face. What would future sages, unaware of the history of our civilization after it mysteriously disappeared, make of these monstrous heads? Would they perhaps think they were representations of gods, which they certainly are not, and thus fall into the same trap we have fallen into regarding many aspects of antiquity? Why and how, they would wonder, did these unknown people carve these great images in a remote place, far from a populated center? Perhaps, at Marca-hausi, there was a far more mundane reason for these sculptures, as there is for ours today.

When these sculptures were new, some ten thousand years ago, it is possible that the climate of the Marcahausi plateau was very different from what it is today. Perhaps it was warm and verdant, just as the present Sahara desert used to be lush grassland swarming with countless animals. Maybe this was merely part of a display in a park which has long since vanished, a park where people could sit and amuse themselves watching the changing lights and shadows alter the figures. Or perhaps it was a display in a zoo, showing the different kinds of animals to be found. In Hamburg there are a group of giant concrete figures representing many species of dinosaurs in the grounds of the zoological gardens; it would be interesting to see what experts from the future would make of these if they survived the ravages of time when all else had vanished.

South America contains hundreds of strange enigmatic carvings and markings, some on a huge scale which have defied all our scientists' efforts to explain. Much has been said in recent years about the maze of lines, trapezoids, and figures on the Plain of Nazca in Peru, which can only be seen properly *from the air*. Von Däniken in *Chariots of the Gods* has suggested that this display is an airfield for spacecraft. He suggests that *originally* there were several landing strips laid out either by alien astronauts or with the work done by the primitive inhabitants under the instructions of the aliens, and that the later lines and figures were added by the terrestrial primitives, either copying what the "gods" had made or in an attempt to signal them.

One feels somewhat uneasy about such a theory. For one thing, it is extremely doubtful if space vehicles, particularly of the advanced design envisaged for long-duration space flight, would need landing strips or runways, as they would more than likely descend and ascend vertically.

The figures of Nazca were only discovered from the air; from the ground all that can be seen are white markings against the dark surface, without any discernible pattern. It is a mystery how these figures could be created without careful measurements, perhaps on a small-scale pattern, and with progress checked *from the air,* since making them with only a surface view would be almost impossible. The figures are on a vast scale, and very accurate; among the designs can be recognized a monkey, a spider, and several representations of what appear to be hummingbirds and a condor.

It is thus possible that even if all the designs were not made at exactly the same time, they were made by the same technique, which would include *aerial survey*. If some of them were made by advanced methods (von Däniken's astronauts) and the others done by primitives, there should be an enormous difference in the design ability, but *all* parts are of equal skill. The author suggested in *Colony: Earth* that the pattern may have possibly been an aerial direction indicator, as it seems to bear a resemblance in some of its aspects (particularly the lines themselves) to a string navigational device used by the Polynesians to enable them to find their way very efficiently across the trackless Pacific.

Indeed, many of the enigmatic markings which exist in South America seem designed to be seen specifically from the air, for either they virtually cannot be seen from the ground, or from surface level they make no sense at all.

For more evidence of ancient flight there is, in the Bay of Paracas on the coast of Peru, the Chandelier of the Miraculous Sign of the Three Crosses. This figure in white on a cliff face is somewhat like the trident of Neptune with branches, and stands 185 feet high. It has been suggested that it was a guide marker for ships, but because of the enclosed nature of the bay it can only be seen properly close inshore, which seems to exclude that possibility. A Spanish scholar, Beltran Garcia, has suggested that by pulleys and cords a pendulum could be created, using the trident, that could record seismic disturbances not only in South America but anywhere on earth. He thus suggests its purpose is as a seismograph. However, the figure can be seen most effectively *from the air,* and perhaps it was really a navigational device of some sort, or even an aerial symbol denoting a national boundary.

Strange markings appear on Andean slopes, resembling parallel lines set closely together, across the hills. Their makers and their purpose are quite unknown and, as with the Nazca lines, can only be detected properly *from the air*.

It is a mystery since South America is full of strange markings and carvings which can only be seen properly from high in the sky, in some way they must be connected with manned flight in a distant epoch. Their true purpose remains as much a mystery to us as many of our symbols of technology would be to a distant future which had forgotten us.

Of course, the traditionalist will say that there could not have been aircraft in the remote past; that these carvings were made by primitive peoples to appease their celestial gods. But no one has suggested how things like this were created without high-altitude surveys, and no one today would attempt to reproduce such figures from ground level with primitive tools and by rule of thumb, because it simply cannot be done!

In the same way, many geologists will say that the light effects on the carvings at Marcahausi occur because of weathering during the past few thousand years, and that therefore the effects are accidental. Yet the same effects are seen on stone in Britain and Europe and, according to Dr. Ruzo, in many other parts of the world, including Egypt. Surely the effects of different climates, different kinds of erosion from wind or sand or water, depending on locality, would not have produced, *accidentally*, identical optical effects. The experts would no doubt deny that ancient sculptures of horses, lions, and so on existed in South America, but there are photographs, *and they are there*, and their existence cannot be denied, however much the traditionalist may wish them away. New explanations are needed—new answers to old problems.

Much of the working of masonry in antiquity, especially in the pyramids of Egypt and the megalithic constructions of the Americas, has long baffled the experts. Enormous slabs of walls from Tiahuanaco in Bolivia seem to have been planed absolutely flat, and grooves carved as though with a knife through soft butter. This has led to the idea that perhaps the Inca (or pre-Inca) peoples had some method of softening stone with a chemical so that it could be easily worked.

In support of this it has been said that there is a plant in the Amazon basin that exudes an acid which is capable of producing this reaction. However, this has never been verified. It is possible that such a plant may exist, but this still leaves other problems unsolved: principally, *how* such huge masses were erected—a question that applies to megalithic building all over the world. It may seem that the *whole* problem of the megaliths could more realistically be solved by assuming the existence in the remote past of a high technology capable of dealing with *all* the aspects of megalithic construction.

In this connection, regarding the actual cutting of the masonry, it is interesting to note that experiments have been carried out recently in the use of lasers for the purpose of cutting stone. These experiments have been proving successful, and, what is more, the effects created are very similar to those on megalithic stones cut in antiquity, even to the glaze produced by the intense heat, which has been observed in some ancient American megalithic structures.

We did in fact suggest in *Colony: Earth* that in view of the fact that no tools have been discovered at many ancient sites, and that the tools which we have assumed to have been used, made of either stone or copper, are ineffective under modern test conditions, something akin to lasers may have been used by the ancients.

We have seen in our chapter on ancient science that many present-day technologies appear to have been known in antiquity, and there is no reason to think that among their devices may not have been something similar to lasers. At least the laser offers a solution to both the apparent ease of stone cutting and the total lack of tool remains. The similarities between the traces on ancient stones and those made on modern masonry by lasers does lend a certain weight to the idea.

Even if saws had been used in antiquity, it is difficult to

imagine the technical difficulties in handling such large slabs by this method, and traces of saw marks should be visible on the stone. No such marks are visible; as Verrill remarked in *America's Ancient Civilizations,* the slabs of stone in the wall masonry at Tiahuanaco seem to have been smoothed absolutely flat, as though with a gigantic plane.

Much comment has been aroused recently regarding the stone balls discovered in Costa Rica, Guatemala, and Mexico. Their sizes range from about eight feet to less than an inch and the largest ones weigh sixteen tons. Some are set in patterns on platforms, and others are arranged in straight lines or in clusters. They are all perfectly round and polished, and they scarcely could have been produced naturally, or we would surely have discovered them in other parts of the world; as it is, they appear to be limited to Central America.

The alternative is that they were man-made objects, and it seems extremely unlikely that primitive people with almost no tools could have produced such perfectly spherical shapes. In many cases, the largest globes would have had to be carved from blocks weighing twenty tons or more, and some must have been transported many miles over difficult terrain to the sites where they have been found.

It has been suggested by Robert Charroux in his book *The Mysterious Unknown* (Neville Spearman, 1972) that when a group of these balls in Guatemala were arranged into what was considered to have been their original pattern, they represented the solar system and the principal constellations. If this were so, it was a remarkable feat on the part of the ancients, for it is within only the last 150 years that we have become aware of the true nature of our planetary system. However, it is not completely impossible, because, as we have seen, there seems to have existed in antiquity a great deal of knowledge regarding the solar system that was later forgotten.

On the other hand, it will be interesting to see what an eventual survey and accurate mapping of all these stone balls may bring to light. It is possible that we may find a parallel between these obviously astronomical arrangements and those of the megalithic observatories in Europe; perhaps there was an equivalent of Stonehenge in the Americas.

Another area of interest is the weaving skill of the ancients. It may seem surprising, but even such relatively fragile things as textiles have survived a great period of time—in some cases, many thousands of years. Even if the actual material has not survived, the impress of its presence has enabled us to reconstruct its appearance and texture. As with other things, we have the extraordinary truth emerging that the more ancient the artifact, the higher its quality. Catal Huyak in Anatolian Turkey is one of the oldest cities known to man, and carpeting has been found there which is of so high a quality that it compares favorably with those made today.

Prehistoric lake villages, which existed in England and parts of Europe, have provided us with evidence of a previously unsuspected sophistication, for traces have been found of a brocaded cloth of a very high standard—not a very easy material to manufacture. It had been thought that these villages were among the earliest settled communities, only one step removed from nomadic savagery, yet here we have evidence that they may have been very far from primitive.

Other mysteries abound—for one example, in southwest Africa there are rock paintings showing Bushmen and white women. One painting, known as the White Lady of Brandberg, shows a white woman with a flower in her hand. Some have thought these figures depicting white people represent Cretans or Egyptians who traveled this far south thousands of years ago (the Pharaoh Necho sent an expedition to circumnavigate Africa), but they more resemble Caspians from North Africa who lived there twelve thousand years ago. Among the animals painted on the Brandberg rocks, leopards and hippopotami are noticeably absent. It *may* be coincidence, but these particular animals did not occupy this region many thousands of years ago, although they do now, which may lend support to the idea of the great antiquity of the drawings.

West of Alice Springs in Australia, Michael Terry found a carving of the extinct *Nototherium mitchelli* on a cliff face. This species vanished some 2,500 years ago from the Australian continent. Not too extraordinary, perhaps—the Australian aborigine has lived in Australia for many thousands of years—but in the same place were found six representations of what appear to be rams' heads. Another most extraordinary drawing was of a man of European features with a beard, drawn in a horizontal position and wearing a miter which closely resembled Egyptian or Babylonian design. Imagine a white man, or someone who could have originated in the Middle East, on a rock face in Australia, where the ram was unknown until introduced by European settlers in recent times! Erosion of the rock faces, which has blurred the carvings, hint at their great age.

Did people from Europe or the Middle East visit both southwest Africa and Australia in antiquity? It is not impossible, especially in view of the Piri Reis map showing Antarctica; if Antarctica was known in ancient times, it is possible that the countries at the "bottom of the world" were also known.

Another unexplained oddity lies in the field of medicine. Perhaps one of the most important aspects of the human condition, of great interest to every human being from time immemorial, is the matter of health. Even if a society had been sufficiently advanced to have eliminated disease, there is still the problem of accidents. Western society has largely controlled epidemics, but the hospitals contain a large proportion of people who have been injured in various kinds of accidents. Therefore it is necessary to have a thorough knowledge of human anatomy, physiology, and surgical techniques to help in the cure of accidents of all kinds.

We have already mentioned in our chapter on ancient science that a great deal of medical knowledge existed in the past, and that such things as vaccination were foreshadowed thousands of years before Jenner and Pasteur. In this connection it is interesting to note a curiously modern approach to surgery in ancient Egypt. We have already mentioned the fact that Egyptian physicians of antiquity appeared to be aware of the circulatory system. Our knowledge of this fascinating anomaly from a remote period of Egyptian history stems from one particular source, known as the Edwin Smith Papyrus.

Edwin Smith, a pioneer American Egyptologist, bought the papyrus from a dealer in Luxor in 1862, but his knowledge of Egyptian was insufficient to translate the technical nature of the manuscript. His heirs gave it to the New York Historical Society, and it was then translated by the famous Egyptologist J. H. Breasted, with the aid of the distinguished physician Arno B. Luckhardt.

The papyrus dates back to the Egyptian Old Kingdom (c. 3000 B.C.), although Egyptian civilization is reckoned to be older than this. It can be assumed from the nature of the text of the Smith Papyrus that it is a copy of the original, and it appears that the writer was not overly familiar with some of the technical terms he was copying (or translating?), as he interposed explanations. This leads to the assumption that the copyist was not one versed in medical or surgical techniques.

The papyrus is, however, a remarkable document in that, unlike many Egyptian medical texts that have survived, it is completely devoid of supernatural overtones and religio-magical practices, being thoroughly rational and scientific.

The Smith Papyrus was apparently intended to be a complete review of surgical techniques, and it may have constituted a small fraction of a complete encyclopedia of medicine and surgery. However, the part we have in our possession is limited to surgical matters connected with the head, arms, and chest. It lists 46 surgical cases, mostly injuries: 27 of the head, six were injuries of the throat and neck; two were of damage to the clavicle; three were injuries to the humerus, and eight were cases of the chest, including tumors and abscesses.

The injuries were described in detail, each with a heading

which began with "Instructions," followed by (for example) "concerning a gaping wound in his head, penetrating to the bone and splitting his skull." The "Examination" detailed the interrogation of the patient, inspection, palpation, and the execution of movements under the direction of the surgeon. Under "Diagnosis" the physician listed the three possibilities open to him: "An ailment which I will treat" (curable); "An ailment with which I will contend" (possibly curable, perhaps fatal, but treatment would be attempted); and "An ailment not to be treated" (a serious case with a possibly fatal outcome for which the surgeon would not accept the responsibility).

This papyrus is the first document in recorded history to mention the word "brain," and its convolutions were likened to the corrugations formed on cooling slag from molten copper. The pulsations of the brain under the surgeon's exploring fingers when it was exposed are compared to "the throbbings and flutterings under the fingers, like the weak place of an infant's crown before it becomes whole." The meanings and the contained cerebrospinal fluid, the sutures of the skull and other skeletal details were mentioned and described in clinical fashion.

The physicians who originated the papyrus were aware of the pulse and its significance in indicating the patient's state of health; this observation was "like measuring the ailment of a man." They were aware that the heart was the organ responsible for the pulsating vessels in all parts of the body—something that was not to be known again for another five thousand years, until William Harvey demonstrated the circulatory system in 1628.

The physicians of this remote time had a thoroughly modern approach to brain injuries and their effects. They were aware of paralysis of the extremities caused by injuries to the brain, and knew that the left hemisphere of the brain controlled right-side motor movements and vice versa. Unlike many other ancient cultures, the Egyptians of the Smith Papyrus did not use trepanning in the treatment of skull injuries. Soft-tissue wounds were treated with sutures or adhesive tape; splints and bandages were applied as they are today. It was noted that skull fractures were frequently accompanied by bleeding from the ears and nostrils. In fractures of the midcervical vertebrae, priapism, seminal emission, and involuntary urination were all noted, with emphasis on the extremely unfavorable prognosis.

It has been said that this first recorded surgical treatise still makes very sound reading for modern surgeons.

The Edwin Smith Papyrus is unlike any other medical text in antiquity. The knowledge it holds of the small section of the human anatomy described is so exact that if it is only a part of a larger work, the extent of ancient knowledge on surgical, and possibly also medical, matters may have been as extensive as it is today.

It seems obvious that this knowledge did not spring from the ancient Egyptian civilization but must have been *inherited from an earlier time.* In this respect it has a distinct relationship to the ancient maps, which are themselves fragments of a larger whole and are as advanced as modern charts, and to the Antikythera Computer, which also could not have stemmed from the archaic Greek civilization.

We would suggest that the Smith Papyrus was a surviving fragment of the knowledge contained in the pyramids by an advanced civilization of remote times. Possibly the complete copy was in the great library of Alexandria, and this fragment, by a fluke, had escaped destruction when the library was destroyed.

We have already mentioned that the Greeks obtained most of their knowledge from Egypt, and this is true of medical matters. The Father of Medicine, Hippocrates, was born in 460 B.C., and the more than seventy books called the *Corpus Hippocraticum,* although attributed to Hippocrates, were probably a team effort by his followers and students.

The grasp of surgical matters displayed by Hippocrates is remarkable, and in the matter of dislocations, this knowledge was *more extensive* than that possessed by many modern surgeons, unless they specialize in orthopedics. Special deformities were discussed, and the treatment of clubfoot by nonoperative techniques is described in great detail. Bladder stones and hernias are almost the only common medical problems not mentioned in the Hippocratic texts, and it is possible that studies of these existed but have been lost. Hippocrates and his followers traveled widely and studied wherever they went. Some of their observations, methods of treatment, diagnoses, and prognoses are so similar to the Smith Papyrus that it could be reasonably suggested that much of their knowledge was gathered from the scrolls of ancient wisdom held in the great libraries of Egypt.

There may have been others who benefited from ancient knowledge. Leonardo da Vinci, hailed as the most complete genius in history, a man "five hundred years ahead of his time," could be one example.

Leonardo invented many things: a water-filtration plant, a helicopter, aircraft, novel weapons of war—and in addition was a talented writer and artist. He seemed fascinated by the secrets of flight, as his many extant drawings show. Recently a drawing by da Vinci has been discovered showing in graphic detail a *lunar landing.*

Should we not wonder: where did he obtain the remarkable inspiration for this, and possibly for his other innovations? Was he the matchless genius we have always thought, or did perhaps some of his works not spring solely from the brain and imagination of this man? Did he, perhaps, have access to ancient and secret texts and plans that showed aircraft and space travel? These things are recorded in mythology; perhaps somewhere there existed more concrete evidence than old myths of a vanished age: actual texts, technical information and drawings, plans and designs. Of course, da Vinci could not put any of these things to practical use; the technology just did not exist in his age to construct such things as aircraft.

Possibly some of these documents and drawings may still exist somewhere, hidden away long ago from those who would destroy them and now forgotten. Leonardo da Vinci, if ancient texts were really his inspiration, would not, even if he wanted to, be able to proclaim publicly the source of his genius. Ancient texts, those from pre-Christian times, were in the Dark Ages of Europe hunted down, hidden or burned by the zealots of the Catholic Church. Frequently their owners were likewise hunted and burned as traffickers with the devil.

Much knowledge from a remote and vanished civilization may have survived the wars, ignorance and disorders of the pre-Classical and Classical world, only to have been destroyed by the mindless superstitions of the early Christian world. From the fragments that survive, only now are we capable of realizing the enormity of our loss.

11

The Myths of Our Time

OUR HISTORY has been grossly distorted; to anyone with an open mind this must be obvious. A new appraisal of the past should be made, and it will no doubt be found that our history books, especially those relating to the more remote eras, will have to be rewritten. The same old, hackneyed versions are still being taught in our schools and universities, and new discoveries are ignored or glossed over—a state of affairs that cannot last indefinitely.

Not all the blame can be laid at the door of our teachers, experts, and professors. Too often has a new approach to the past of the human race been dismissed or ignored because the ideas are outrageous or "cranky." This author agrees that much that has been written on the subject is, quite frankly, unrealistic and, at worst, downright nonsense.

Most scholars refuse to associate themselves, for example, with the Atlantis problem, and quite rightly. Possibly no subject on earth has been the basis for so much speculation, theorizing, and plain idiocy. From the writings of Plato, people have expanded Atlantis to a vanished island (or continent) complete with all the gadgetry of superscience and a master race of geniuses. Why this was done is hard to say—possibly much of it is wish fulfillment, and a little may be the recognition that there could have existed

174

in the past a highly advanced civilization which was completely destroyed.

The mistake has been to associate this with Plato's *Atlantis*. Plato, in his dialogues, could not have been clearer. He was describing a civilization of the Aegean world: he mentions chariots, bowmen, spearmen, the style of buildings and temples which were a familiar part of the Mediterranean world for at least two thousand years up to the beginning of the Christian era. Despite this, Atlantis has been placed in nearly every region on the face of the earth, and every archaeological curiosity has been credited to Atlantis and the mysterious Atlanteans.

There are those who will say that Plato reduced Atlantis to terms familiar to him in his day—that he did in fact render a "Greek" version of Atlantis out of the story handed down to him by his ancestor Solon. There is no evidence to support this. However much people want to believe otherwise, I believe the Atlantis legend has been solved: it was the volcanic destruction of Thera (Santorini) and the consequent destruction of the Cretan sea empire in 1500 B.C. Yet the myth is still being perpetuated, based on Plato's *Atlantis*, and we do not need it to solve the mysteries of our past. There are threads that lead to another, infinitely more ancient, civilization than the Atlantis of Plato, but to which we can give no name, for no name has come down to us.

Possibly it never had a name as such, in the same way that our civilization has no name. We have what we describe as our twentieth-century technological civilization; it is not specifically English or American or German or Japanese or Russian. It is a *condition* that has spread over the entire planet.

The same situation may have arisen in the past—the endless duplication of legends from all over the world about a vanished Golden Age points to this. If our civilization were to vanish in a gigantic holocaust, a similar series of legends might arise, and they also would be of worldwide distribution—*because the civilization which gave them birth was worldwide.*

Hand in hand with the Atlantis theories goes the myth of the other supercivilization on the other side of the world: the vanished Pacific continent of Lemuria or Mu.

Lemuria commenced its life as a scientific hypothesis to explain the existence of lemurs, those small animals which are regarded as the most primitive representatives of the primate family. Lemurs live largely in Madagascar, but they are also found in Africa and tropical Asia.

William T. Blandford suggested that there may have been a land bridge connecting southern Africa and India. This idea was taken up by the German biologist Ernst Heinrich Haeckel, who suggested that this land bridge was the method by which the lemurs populated the various continents during the Cenozoic (Age of Mammals) which began some 70 million years ago.

This concept of a land bridge connecting continents was seized upon by some people, especially the occultists, and was expanded into a hypothesis of a Pacific continent populated by an advanced race. Incidentally, the original name, Lemuria, is simply derived from the connection with the lemurs.

The most famous exponent of the lost continent of Lemuria was the founder of the Theosophist movement, Madame Blavatsky, who used it in her weird cosmos. Into it she also incorporated Atlantis, an imaginary continent based on ancient Greek allusions to the Land of the Hyperboreans, which she called Hyperborea, situated in the Arctic.

Mu is derived from Churchward, who thought that the Maya civilization in Yucatán was a fragment of a sunken continent called Mu. The name Mu is based on a supposed translation of two Maya glyphs regarding a tradition about the Flood, but as no Maya glyphs apart from numerical ones have ever been translated, there is no certainty that the Maya ever actually called anyplace "Mu." The *Popol Vuh does* refer to the Flood and the destruction of the first men, and the Mexicans refer also to a great catastrophe and the Dead Lands to the North. However, neither makes reference to sunken continents as such.

The geography of the Pacific would seem to exclude the possibility that there ever existed a continental mass which sank within human times, although possibly there may have existed different land areas many hundreds of millions of years ago. The same thing applies to the Atlantic Ocean, where both oceanographers and geologists are agreed that no large land area could have existed within geologically recent times.

The discovery of land-created lava from the bed of the Atlantic has been sometimes cited as proof that there did exist a continent or huge island in the Atlantic which could have sunk. Of course, it is possible that a *small* volcanic island did vanish beneath the waves. However, it is exceedingly doubtful that a large land mass could have disappeared beneath the Atlantic, particularly within the last ten thousand years, without leaving noticeable traces for the geologists.

Although the Atlantis theory is still strongly advocated, with new locations being constantly advanced, the hypothesis of a Pacific continent has largely fallen from favor. The occasional arguments put forward in its favor center largely around the mysterious giant statues of Easter Island and the lagoon city of Nan Matol on Ponape in the Carolines. Here, the pro-Mu or -Lemuria theorists say, is proof of a vanished continent. These mysteries, they claim, stem from Mu (or Lemuria), being the last visible remnants—the part that did not disappear beneath the waves.

Easter Island, it is true, does pose a riddle which has never been resolved. Apart from the enormous number of huge statues (the great heads actually have bodies, rather small in proportion to the heads, buried in the ground), there have been found a number of tablets inscribed with symbols that appear to be some form of writing. The inscriptions on these tablets have never been deciphered.

Thor Heyerdahl's explanation of the statues seems the most reasonable: that they have a connection with one of the cultures of the Peruvian/Bolivian region. There do exist certain similarities between them and the megalithic statues of Tiahuanaco on the shores of Lake Titicaca in Bolivia. However, the faces at Tiahuanaco are completely different from those of Easter Island—the heads of Tiahuanaco are square and stylized, while those of Easter Island are long, with sunken eye sockets and long noses. Similarities also exist: both groups wear stone "hats" and the Easter Island statues have long ears, which have a similarity to the "Long Ears" the Spaniards noted in the ruling elite of the Inca of Peru. No writing in South America has ever been found which is comparable to the Easter Island script, although here we should note that no direct connection has been found between the Easter Island script and the statues. Those who carved the statues may not have written the tablets.

Apart from Thor Heyerdahl's, no satisfactory archaeological explanation has been offered for the creation of the vast number of huge statues on Easter Island, which is small and bleak and cannot support a large population. One would have thought that an army of workmen would have been needed to carve and move these statues from their quarries to the hillsides where they were set up, to say nothing of all the ancillary people such as food suppliers, house builders, and the attendant women and children. Easter Island never could have supported such numbers.

Erich von Däniken has suggested that spacemen were marooned on the island and built all these statues with the aid of their advanced technology, either as a signal or simply to while away the time awaiting rescue. This would account, he says, for the number of unfinished statues which are lying there: as soon as the rescuers came, the spacemen dropped what they were doing and went away into the sky. Hence the island's other name, "The Island of the Birdmen." Incidentally, it has still another Polynesian name: "The Navel of the World."

It does seem unlikely, not to say ludicrous, that a group of spacemen marooned temporarily on an alien planet would spend their time building a vast number of great statues, all alike, and setting them up over the hillsides. It seems very doubtful if representatives from an advanced technological civilization would go to such lengths merely to pass the time. As for signals, they would probably have more sophisticated methods, such as a directional radio beacon.

The mystery of Easter Island, therefore, remains such, and is likely to cause a great deal of controversy for many years to come.

Considering another Atlantis myth, we turn to Nan Matol on Ponape, of the Caroline Islands, an even more unlikely candidate as the remains of a highly advanced civilization which was left high and dry when all the rest sank.

There is a broad bay in the southeastern end of Ponape, and the ruins of Nan Matol are situated on a small island called Temuen which, at high tide, is broken up into almost a hundred tiny islets. Most of the islets are surrounded by huge walls, thirty feet high. At high tide the eleven square miles of Nan Matol appear like some ruined Venice.

The walls and all the buildings are built of a dark-blue prismatic basalt, a rock similar to that which forms the Giant's Causeway in Ireland. It appears that the builders of Nan Matol obtained their materials from the island of Jokaz off the northern coast of Ponape, and rafted them fifteen miles around the island to the site. Stones scattered along the sea bottom on the route show where some of these rafts must have sunken with their loads.

The six-sided lengths of prismatic basalt were crudely laid in alternate courses in the wall, rather after the fashion of a log cabin. There are numerous holes, and no attempt had been made to dress the stone or ensure a close-fitting coherent mass.

The ruins are impressive, especially when seen from a distance, but it could not possibly be suggested that they were the handiwork of an advanced civilization. No statuary has been found, and no inscriptions of any sort have ever been discovered within the precincts of the complex. After careful sifting through the legends of the Ponapeans, together with analysis of radiocarbon methods, it is now considered that Nan Matol was built in approximately A.D. 1400—or rather from then onward, being added to by successive kings with the title of Satalur.

There are the remains of the kings' house, priests' quarters, temples, and altars, so the place was apparently a cult center connected with the worship of the Sacred Turtle. There is no possibility that this was a remnant of a highly advanced civilization called Lemuria which sank beneath the Pacific. Even so, no doubt someone at some time or other will attempt to revive the theory of a sunken continent in the Pacific, in the same way that the Atlantis myth is being constantly resurrected in differing forms.

For example, many people have reported that they have seen the ruins or outlines of buildings in shallow coastal waters. An aircraft observer during the war said he had seen huge buildings in shallow water off the Pacific coast of South America. Divers have claimed to have observed sunken buildings in the Gulf of Mexico, and there are reports that the bed of the Arctic Ocean is littered with the remains of a submerged forest.

It has been said that during the last Ice Age so much water may have been locked within the ice fields that the ocean levels may have been lower than they are today, and that extensive areas of the continental shelves of many lands could have been inhabited. As the ice melted these areas were flooded when the oceans rose to their present levels, and the inhabited areas were abandoned. This is one explanation for the possible existence of ruins in shallow offshore waters.

Alternatively, as was suggested in *Colony: Earth*, there may have been no Ice Age, but rather a warmer climate. This would still have meant a lowering of the ocean levels as a result of a higher rate of evaporation of surface water. The sudden lowering of temperatures would have caused excessive precipitation, creating the same effect as the melting of the hypothetical ice fields—a rise in ocean levels.

However, even this hypothesis is open to question, and it is possible that there was actually no lowering of ocean levels within human times. This may be borne out by making reference to the ancient maps previously referred to. We observe that these maps show both the South and North Polar regions without ice. Yet the maps show all the continental outlines *as they are today*, and the distances between continental masses are the same as today, and very accurately charted. They therefore appear to show a world devoid of ice, in opposition to the Ice Age theory, yet they do not show any lowering of ocean levels that would expose any of the continental shelves.

Therefore, even if the planet had been warmer, with increased evaporation, there could also have been increased precipitation —in other words, a warmer but perhaps slightly wetter world, which would have maintained the ocean levels at a level similar to today's. The fact that many thousands of years ago all our deserts were fertile areas would seem to point to more rain in the past.

How, then, do we explain the buildings seen—or claimed to have been seen—in offshore waters? There are two explanations: one is that they do not exist, and what has been seen are natural formations. The Giant's Causeway is a natural formation, but the rock formations are so regular as to almost appear to be artificial. Submerged prismatic basaltic formations could be mistaken for artificial constructions.

Or secondly, the remains of buildings and the trunks of trees that litter the bed of the Arctic Ocean *could* have found their way to these locations during the enormous disturbances created by the Flood catastrophe. Tremendous gales and tidal waves could have torn off coastal buildings and vegetation and carried them to shallow waters surrounding the coasts.

Of course, there is always the possibility that there were buildings in what is now coastal waters. It may also be true that the Arctic and North Sea were once very dry land and that England was joined to the continent of Europe. We then would have to explain the ancient maps of Piri Reis; they show modern continental outlines with great accuracy, including ice-free polar regions, yet the ocean levels appear to be the same as today's.

Either this was so and England was not joined to Europe in human times, or the maps were drawn up to show the dispositions of land and sea after the catastrophe, but before the ice caps had formed in their present locations. Alternatively, the maps may have been based on charts drawn when the ocean levels were lower, but altered to show the newer levels as navigational aids in the period following the changed state of the world.

On all the evidence, however, it is beginning to appear that there may never have been any sunken lands occupied by the human race—not of any consequence, at any rate, and not during the past thirty or forty thousand years.

The unknown civilization that vanished before the dawn of recorded history may in fact have occupied all the areas we do today, even including ice-free polar regions at present virtually uninhabitable.

We have attempted to show in this volume that there may have existed in the remote past a highly advanced civilization that was destroyed in a vast catastrophe, possibly intelligently engineered. There exists the distinct possibility of a nuclear holocaust in the past. This disaster may have been limited to this planet, or it may have involved three or possibly four inhabited planets within our solar system, one of which was completely shattered by weapons of frightful power and whose fragments scarred the other inner planets.

Both the nuclear weapons used on earth and the bombardment of fragments of a shattered world may have been responsible for the destruction of a civilization and a shift in the earth's axial and orbital position.

There is a possibility that there used to be a planet where the asteroid belt is now situated. The scarred condition of the inner planets and the moon, the existence of glass "marbles" and glassy areas on the moon, and tektites on earth, which could have been formed by the intense flash heat of nuclear reactions, all point to the possibility of a sustained artificially engineered chain reaction that destroyed a planet.

Such a possibility would fit in with some of the Old Testament statements about the heavens being shaken and the earth being moved out of its place; the theory would also fit some sections of the Apocalypse in Revelation, which describe the War in Heaven. In one place it is said that a great star fell to earth.

It would appear that much of Revelation is concerned with destruction. However, a great part of the writing, and the psychological background that underlies it, is connected with the tremendous Thera volcanic eruption, which caused such havoc in the second millennium B.C. and left a traumatic shock of such magnitude among all the peoples in the crowded eastern Mediterranean that it took them centuries to recover. However, there are also echoes of earlier disasters and conflicts. Those would seem to have been telescoped into a generalized series of disasters which have to a great extent been stylized into a religiously oriented Nemesis.

The Bible, particularly the Old Testament, has been subjected to a great deal of scrutiny in recent years, and it is highly popular at the present time to attempt to identify a great deal of biblical phenomena with the activities of visitors from other worlds. It may be that there is a danger of giving too much weight to such ideas; many parts of the Old Testament are so ambiguously written that almost anything can be translated or interpreted from some statements.

There is no doubt that some things in the Old Testament can be equated with the phenomena we today describe as UFOs —particularly the episode of Moses on the Mount (although as Exodus is so closely linked with the Thera eruption this could also be connected with vulcanism). The episode of Ezekial also has a strong resemblence to present-day UFO reports.

However, a great many gaudy, unreal, and impossible cosmic fantasies have been evolved from biblical passages, and around the legends about Atlantis, Lemuria, and Hyperborea. How certain modern writers can create a complete continent populated by advanced civilizations out of almost no evidence is incredible.

For example, the Greeks mentioned an island called Hyperborea, "beyond where the north wind blows." On it was a round temple dedicated to Apollo, and the inhabitants were favorably disposed toward the Greeks.

There is not the slightest necessity to invent an entire continent out of a Greek legend. Like most legends, it has within it a core of truth for which a quite reasonable hypothesis can be made without inventing another vanished land.

The region referred to is obviously somewhere far north of Greece, which could mean the British Isles or Scandinavia. However, the fact that Hyperborea is mentioned as an island where there was a round temple would seem to narrow it down to the British Isles and Stonehenge (the Round Temple).

The reference to Apollo is interesting, for it may be that there is a distinct connection between Apollo of the Greeks and the basis of an ancient name for Britain: Merlin's Enclosure. There has arisen a degree of confusion about the name Merlin, the magician of King Arthur's court who has been associated with ancient monuments such as Stonehenge. It is thought by some mythographers that the name Merlin's Enclosure is a corruption of an even more ancient name; a Welsh triad says that before men came to the British Islands it was called "Clas Myrddin"—and Myrddin was an ancient Celtic "sky god."

Apollo, as a celestial deity of the Greeks, could be related to the Celtic Myrddin. It was said in Greek mythology that Leto, the mother of Apollo, was born on the island of the Hyperboreans, and the priests of the island were regarded (by the Greeks) as priests of Apollo. One is tempted to wonder whether the relationship between Myrddin and Apollo was even closer: could they in fact be aspects, under different names, of the same celestial deity?

It is also interesting to note that the Greeks said that on visits to the Hyperboreans they left gifts and votive offerings inscribed in Greek, and at Stonehenge there are representations of Greek swords and Greek lettering carved on the columns. This has led to the theory that Greek mathematicians or traveling architects and builders were responsible for the construction of Stonehenge; now it begins to appear that it was actually the other way around—that in fact the Greeks came to Stonehenge to *learn* from the builders, the mysterious Hyperboreans.

We remember at this point the Antikythera Computer, and our suggestion that this may have originated from the Stonehenge mathematicians in Britain. This not only strengthens the Greek ties with the Hyperboreans, but makes the Greeks the students and the Hyperboreans the builders and teachers.

It must be obvious that a close study of the Greek legends regarding the land of the Hyperboreans ties it more closely to

Britain than to any other area. All the facts, both from a mythological point of view and from the discoveries of actual physical
remains, particularly in the area of Stonehenge, would seem to
suggest that the island of the Hyperboreans was actually Britain.
The invention of a sunken or otherwise vanished mythical continent situated in the present polar regions is not only unnecessary
but absurd.

It is remarkable how a vast *modern mythology* has arisen out of
two Greek legends, both obviously based on fact. Atlantis and
Hyperborea have been the subjects of endless series of speculations and theories, and each has been used to explain a great many
of the mysteries of the past.

Much of the blame for the modern mythology regarding Atlantis, Lemuria, and Hyperborea can be placed at the door of
someone we have already mentioned: Madame Blavatsky. Although she has been mostly forgotten by the public at large and
her rambling, nonsensical universe dismissed for the fabrication it
obviously is, her influence still remains strong among certain of
the modern writers engaged in speculation about the past.

Briefly, the Blavatsky universe, incorporated in her monumental Theosophist work *The Secret Doctrine,* is made up as
follows:

The universe was shown to her in a trance by means of an
Atlantean "history," the Book of Dyzan. There were different
"Root Races," the first being a sort of invisible jellyfish who lived
in the Imperishable Sacred Land. The Second Root Race lived in
Hyperborea, which was situated in the Arctic and broke up.
Following the destruction of Hyperborea, Lemuria formed in the
Southern Hemisphere and was inhabited by the Third Root Race,
giant apelike creatures with four arms and sometimes with eyes in
the backs of their heads.

Lemuria in turn sank beneath the sea, and its place was taken
(in the North Atlantic, however), by Atlantis, inhabited by the
highly advanced Atlantans, the Fourth Root Race. We today,
following the destruction of Atlantis and the development of the
present land masses, are the Fifth Root Race. It appears that the

Sixth Root Race, to follow us, will evolve from present-day North America, and the Seventh (and final) Root Race will emerge from South America.

One wonders why she did not invent some new continent to emerge like some watery phoenix from the depths of the ocean!

This farrago of ill-conceived nonsense was based on several things: the Greek writings of Atlantis, extant in the *Criteas* of Plato; the widely known legend of the island of the Hyperboreans; Donnelly's monumental *Atlantis,* and the many occult works of the late nineteenth century. The Book of Dyzan was cribbed from the *Hymn of Creation* in the Sanskrit *Rig-Veda,* Lemuria from the hypothetical land bridge, and Churchward's Mu of the Mayas.

It is difficult to understand how anyone could be taken in by this patently unreal "history" of the world, yet some modern writers appear to have rediscovered *The Secret Doctrine,* and have evolved new theories based on the Blavatsky thesis.

Robert Charroux, a French writer, seems to have based many of his ideas on the writings of Blavatsky, particularly *The Secret Doctrine.* He has also been influenced by a curious so-called legend regarding the founding of the city of Tiahuanaco in Bolivia. This legend has it that Tiahuanaco was founded by a woman called Orejona, who landed in a golden spaceship near Lake Titicaca and gave birth to the human race by mating with (of all things) a tapir!

This particular legend is not one that figures in any of the traditions of the Andean peoples, either Inca or any of their forerunners. It was reported by Bertan Garcia, who claims he saw it in a secret manuscript belonging to the historian Garcilaso de la Vega—a manuscript no one else has ever seen. The legend would seem, therefore, to be the work of someone's imagination during the sixteenth century.

As Peter Kolosimo says in his book *Not of This World,* Chapter Two, "The Devils from Space":

"It is depressing to see how Charroux, a writer who is certainly not rigidly scientific but at least appreciated by some for his brilliant deductions, has sunk down to the histrionic level of

Adamski. And it is still more melancholy to note that this is end of many an investigator who, having seriously approached unusual problems, falls for the charms of crude theories, queer associations of ideas and interpretations. Thus they compromise themselves as they are unable to withdraw from the positions they take up and end by having to resort to distortions and falsehoods.

"Apart from discrediting themselves, they obviously increase the destructive and slanderous effect which the champions of orthodox scientific conservatism have on the genuine students who are engaged in revolutionary research."

This writer agrees wholeheartedly with that sentiment, which he expressed earlier in this book to the effect that the "lunatic fringe" automatically brings scorn on the heads of those who suggest an approach different from the orthodox school of thought.

Charroux, in his distortion of the Blavatsky doctrine, sees the descendants of the vanished continent of Hyperborea as being the Celts, who were responsible for the civilizations of the Mexicans and Central American races. He claims, in effect, that the ancestors of the Aztecs and Mayas were Celts from Northern Europe, who are in fact latter-day Hyperboreans. A similar line of reasoning was adopted by Beaumont in his book *The Riddle of Prehistoric Britain*, where he too stated that in his opinion the Aztecs were descended from Celtic immigrants in prehistory.

To a somewhat lesser extent, von Däniken (*Chariots of the Gods*) appears to have been influenced by *The Secret Doctrine*, as he quotes this as one of his sources and also appears to take seriously some of the peculiar ideas expressed there.

What is disturbing is the thought that the day may come when any one of these curious theories about sunken continents could be given a ring of authenticity by an established orthodox figure, any one of whom *may* decide at some future time to support the contention that such a continent existed. Perhaps, in some time to come, it may be written into the textbooks that there did exist a continent in the North Atlantic called Atlantis from which the civilizations of the Old and New Worlds were descended. Selective evidence, especially in view of the many new discoveries that have been made since the end of the Second World War, could advance a very plausible case for the existence of such a continent. It could be argued that it is the most likely theory to fit the facts to hand.

This is not fantasy, nor is it completely unlikely; it has happened before on more than one occasion.

We have the theory of evolution, about which so much has been written and which is propagated with such enthusiasm in every school and university *in the world* that most people believe it is an actual, demonstrable *fact*. This it is certainly not. It is a *theory*, created by orthodox science to oppose the (to the scientific mind) unacceptable concept of divine creation. The theory is composed of scattered evidence, most of which is not very strong, and a mass of often contradictory suppositions. One famous biologist has said that "we believe in evolution, not so much because it is true, but because it is the only alternative we have to divine creation."

We also have the theory of the Ice Ages, again about which countless volumes have been written and which is likewise taught in every school and university. Almost everyone believes that to be absolute truth, demonstrably so in every aspect. Again, it is merely a theory, created to explain certain biological and geographical peculiarities for which there are alternative explanations that are equally valid. The causes of the Ice Ages are vaguely stated, hedged about with innumerable difficulties. For the warmer conditions of the Climatic Optimum, which was supposed to have followed the last Pleistocene Ice Age, there does not exist even a hint of a theory or explanation.

These two accepted concepts are based on evidence almost as flimsy as the evidence for the lost Atlantis or the Lemurian Land Bridge.

It is stated here that we do not need any of these theories to explain the development of our planet and its inhabitants. We need neither a theory of human evolution, nor an Ice Age, nor any number of mythical sunken continents.

We have suggested that this planet could have been colonized from other solar systems. This is not impossible; no doubt we ourselves will do that some time in the future. What we will one day be capable of doing, other races before us may have done. More than one planet in this system may have been inhabited in times long past; the colonists from other planetary systems may have developed powerful civilizations with space travel and the nuclear apparatus of their own destruction. Possibly this event occurred in the unwritten past, leaving legends of catastrophic events both on earth and in the heavens that would explain many strange aspects of mythology, such as the anger of God and the War in Heaven.

Such a war, waged on an interplanetary scale, will, within the foreseeable future, be possible. What may our remote, primitive descendants make of the stories of such possible events? Will they not perhaps devise a mythology that closely resembles the one we ourselves possess today?

Such a conflict in the remote past would not only explain the curious legends which exist now; it would also explain many other aspects of our past.

It would explain the existence of areas of scientific knowledge which must have stemmed from an era of tehcnology and instrumentation.

It would also explain the sudden emergence of civilization around 4000 B.C. Civilizations in many parts of the world appear to have sprung into existence fully formed, and this points to the survival of knowledge from an earlier time.

We do lack, at the present time, *absolute* proof that such a superior civilization existed in the past. Why, we are asked, if such a civilization existed, have we not by now discovered the remains of some machine, parts of a computer, or an aircraft, or even the rusted remains of a rifle?

There are two principal answers to this question. One is that an artifact of this nature may one day be found; we have simply not found such a thing as yet. On the other hand, the science of a past civilization may have developed along such different paths that we would not recognize such artifacts from this civilization even if we found them.

One could consider the case of the rocks found in woods in Virginia around which nothing will grow. It is a curious thing, but no trees have ever been known to grow in the vicinity of Stonehenge, and in the fifty-six "Aubrey Holes" that comprise the outer perimeter of the monument there are small pieces of bluestone. The purpose of these small fragments is completely unknown. What if these stones had been impregnated with some substance, or had their molecular structure altered in some way, perhaps to emit a certain spectrum of radiation we cannot detect, or a certain wavelength of vibration, which inhibits plant growth?

What we would then have is an artifact: something that has been intelligently altered to fulfill a specific purpose. Any object thus functioning is, properly speaking, a machine. Such rocks, therefore, fulfill the purpose of machines which we are completely unable to recognize. Our common definition of a machine is an artifact of machined parts, operated from a linked power source and manifestly artificial in its appearance. The fact that an object which fulfills the purpose of a machine yet looks like a piece of rough-hewn stone does not make it any less a machine. It is merely that our definition of technology and that of a civilization which had developed along different paths may be fundamentally different.

The other factor is that machines, as we understand the term, are extremely unlikely to survive for thousands of years. It is very doubtful if any of our machines—an automobile, say, or an aircraft or a TV set—would survive for very long in a world totally reverted to savagery. No doubt those things that did not rust away and dissolve back into the ground would be broken up and their metals melted down for use as weapons or crude tools.

There is a case in point: In the grave of a Chinese general— Chow Chu (A.D. 265–316)—there was discovered a metal girdle made of aluminum. This metal is extremely difficult to refine and process from its ore, bauxite, and requires the resources of a complex technology. It would appear that such technology did not exist in China in Chow Chu's time. We remember that in Chinese legends there are reports of aircraft, as there were in old Indian sagas. What if this belt were refashioned from the parts of one of such ancient aircraft, perhaps found rotting away, its true function unrecognizable and unknown? It is not entirely impossible.

More delicate artifacts, such as the fine wiring and circuitry of something akin to our own electronics, would be unlikely to survive for many years, let alone the centuries or millennia that have passed since our hypothetical civilization passed away.

No doubt future sages would be hard put to defend their theories of an advanced civilization during our period, in view of the paucity of physical remains. It is extremely doubtful if any of our artifacts of an advanced nature would survive a disastrous nuclear holocaust and thousands of years of savagery, from both men and the elements.

Yet one thing is certain: The presently held view of the origin and development of civilization, and even of man himself, does not ring entirely true. There are too many anomalies, too many things that should not exist, both in our mythologies and in areas of knowledge, if man had only advanced from barbarism during the past few thousand years. There are also many physical remains that point to a different state of affairs than those which have always been taught: the pyramids; Stonehenge and the other megaliths; the megalithic walls of Peru; and, at the other end of the scale, minute artifacts that seemingly could not have been manufactured without the aid of sophisticated techniques.

The future will show which point of view is correct. One day there will be discovered a document or an artifact which will prove, beyond a doubt, that mankind and all his works were once almost completely destroyed in a vast holocaust. Until that day arrives we must open our minds, formulate bolder theories, and continue the search. The answers we seek may be buried in the ground under our feet, or far away among the beckoning stars.